Automotive Fuels and Fuel Systems

OTHER TITLES IN THIS SERIES

T. K. Garrett AUTOMOTIVE FUELS AND FUEL SYSTEMS,
        Vol. 1: Gasoline
ISBN 0 7273-0114-4 (1991)

C. O. Nwagboso (Ed.) ROAD VEHICLE AUTOMATION
ISBN 0 7273-1806-3 (1993)

D. Bastow/G. Howard CAR SUSPENSION AND HANDLING,
        3rd edn.
ISBN 0 7273-0318-X (1993)

R. Wood AUTOMOTIVE ENGINEERING PLASTICS
ISBN 0 7273-0115-2 (1991)

A. K. Baker VEHICLE BRAKING
ISBN 0 7273-2202-8 (1985)

A. J. Scibor-Rylski/D. M. Sykes ROAD VEHICLE AERODYNAMICS,
        2nd edn.
ISBN 0 7273-1805-5 (1984)

AUTOMOTIVE FUELS AND FUEL SYSTEMS

Fuels, Tanks, Fuel Delivery, Metering,
Air Charge Augmentation, Mixing,
Combustion and Environmental
Considerations

Volume 2: Diesel

T.K. Garrett
C.Eng, F.I.Mech.E, M.R.Ae.S

PENTECH PRESS
London

SOCIETY OF AUTOMOTIVE ENGINEERS, INC.
Warrendale, PA

First published 1994 by Pentech Press Limited
Graham Lodge, Graham Road, London NW4 3DG, England
and
Society of Automotive Engineers, Inc.
400 Commonwealth Drive, Warrendale, PA 15096-0001, USA

© T. K. Garrett, 1994

This publication may not be reproduced, stored in a retrieval system, or transmitted in whole or in part, in any form or by any means, electronic, mechanical, photocopying, recording, or otherwise, without the prior written permission of the copyright holder.

British Library Cataloguing in Publication Data
Garrett, T.K. (Thomas Kenneth)
  Automotive fuels and fuel systems
  Vol. 2. Diesel
  1. Motor vehicles. Fuels
  I. Title
  629.253

ISBN 0-7273-0117-9

SAE Order Number R-134
SAE ISBN 1-56091-510-2

Printed and bound in Great Britain by Bookcraft (Bath) Ltd.

PREFACE

Progressively increasing stringent legislation relating to automotive diesel engines has necessitated some major rethinking on both engine design and a wide range of aspects of the properties of fuels, their quality, in-vehicle conditioning, metering, and mixing and burning in the combustion chamber. Among the spin-offs from the research have been significant gains in thermal efficiency, which have arisen not only from improvements in combustion but also from turbocharging and aftercooling and electronic control of metering and injection. Each of these subjects is covered in depth, all occupying whole or major parts of chapters.

Structurally, Vol. 2 has been developed along lines similar to those of Vol. 1. However, aspects such as, for example, fuel chemistry, tanks and low pressure pipes have been covered in detail in Vol. 1. Consequently, in this the second volume, space has been available for comprehensive discussion of several matters that could not be included in the first. They include natural gas as an alternative to diesel fuel for commercial vehicle power units and, more importantly, methods of increasing the air charge in the cylinders. Among these methods are variable valve timing, induction pipe tuning and turbocharging.

Tracing the processing, handling and combustion of the fuel right through from the crude oil to the point at which the products of its combustion are discharged from the exhaust, both volumes have been written primarily to serve together as a work of reference for practising engineers in the automotive industry and its suppliers, including those who are responsible for the quality of the fuel. However, they will also serve as a comprehensive source of information for students.

Acknowledgements

So many manufacturers and other organisations have very kindly supplied information and illustrations for this book that it is unfortunately impracticable to mention every one of them here, but their names have of course been mentioned in the sections dealing with their products and services. The Author wishes to express his sincere gratitude to them all.

Among those who have been particularly helpful are, in alphabetical order: AC Delco, Allied Signal Automotive, AVL, British Gas, BBC Brown Boveri, Bruntell, Bundy Europe, Castrol, Cummins Engines, ERA, Ford Motor Company, General Motors, Hartridge, Holset, Honda, Khünle, Kopp und Krausch, Lucas, Mechadyne, Mercedes-Benz, Penske, Peugeot, Pierburg, Ricardo Consulting Engineers, Robert Bosch, Rover Group, Scania, Signal Instrument Co, SMM&T, Stanadyne, Toyota, and Volvo.

Contents

CHAPTER 1   CRUDE OIL DERIVED DIESEL FUEL   **1**

1.1	The diesel powered car	2
1.2	Diesel fuel	3
1.3	Properties required for diesel fuels	5
1.4	Cetane No, cetane index, and diesel index	5
1.5	Tendency to deposit wax	7
1.6	Density	8
1.7	Volatility	8
1.8	Viscosity	9
1.9	Effects of different constituents	10

CHAPTER 2   FUEL QUALITY AND ADDITIVES   **13**

2.1	Matching additives to fuels	13
2.2	Additives in premium grade fuels	14
2.3	Cetane No. and combustion improvers	15
2.4	The Wax Problem	18
2.5	Cold weather additives	18
2.6	Dispersants and corrosion inhibitors	20
2.7	Detergents and anti-corrosion additives	20
2.8	Anti-foamants and re-oderants	23

CHAPTER 3   ALTERNATIVE FUELS   **25**

3.1	Alcohol fuels and blends for diesel engines	26
3.2	Prospects for methanol	27
3.3	Driveability	28
3.4	Gaseous alternatives	28
3.5	LPG and LNG	29
3.6	ANG, low pressure containment of natural gas	29
3.7	Compressed and adsorbed natural gas (CNG and ANG)	30
3.8	Carbon, its potential and properties	31
3.9	Safety of CNG	32
3.10	Other advantages of natural gas	33
3.11	Operational experience with natural gas	34

CHAPTER 4 TRANSFERRING THE FUEL FROM TANK TO ENGINE — 36

4.1	High pressure fuel pipes	36
4.2	Pipe end fittings and connections	39
4.3	Fuel conditioning	41
4.4	Solid particles and sludge	44
4.5	Removal of water	45
4.6	Removal of wax	48
4.7	Filtration arrangements	49
4.8	Some production filters and water separators	49
4.9	Fuel heating for operation in very cold climates	53
4.10	Fuel feed pumps	54
4.11	AC Unitac diaphragm type pump	55
4.12	Bosch plunger type pumps	55
4.13	AC Alpha mechanical pump	57
4.14	AC Universal Electric Solenoid fuel pump	59

CHAPTER 5 INJECTORS, AND IN-LINE AND UNIT INJECTION PUMPS — 61

5.1	Injectors	62
5.2	The pintle type injector	63
5.3	The hole type injector	65
5.4	High and low spring injectors	67
5.5	Two-stage injection nozzles	67
5.6	The Stanadyne Pencil injector	70
5.7	Types of fuel injection pump	73
5.8	In-line injection pumps	74
5.9	Fuel delivery characteristics and injection lag	75
5.10	Pressure waves and cavitation	78
5.11	In-line pumps — fuel metering	78
5.12	Lucas Minimec, Majormec and Maximec pumps	80
5.13	Pumping element operation	83
5.14	Some plunger arrangements on other pumps	87
5.15	Engine control	88
5.16	Governing and governors	88
5.17	Governor mechanisms	90
5.18	Mechanical governors produced by Lucas	94
5.19	Bosch in-line pumps	96
5.20	Bosch electronic control	98
5.21	Unit injection	101
5.22	Electronic unit injection — Lucas EUI system	104
5.23	The GM unit injector	108
5.24	Penske/Detroit Diesel electronically actuated unit injection	110
5.25	Cummins PT unit injection system	113
5.26	Cummins PT combined pump and injector units	115

CHAPTER 6 ROTARY AND DISTRIBUTOR TYPE INJECTION PUMPS — 118

6.1	The Lucas DP series distributor pumps	118
6.2	The DPA pump	119
6.3	Governing	123
6.4	The Lucas DPS pump with torque and boost control	125
6.5	DPS fuel supply system	127
6.6	Engine starting — DPS latch and rotor vent valves	130
6.7	Limiting maximum fuel delivery	131
6.8	The two-speed governor	134
6.9	Scroll plates and boost control	135
6.10	Automatic advance and retard unit	136
6.11	The Lucas DPC pump	138
6.12	External control of low-load advance	145
6.13	Electronic control of distributor pumps — Lucas EPIC system	150
6.14	Application and benefits of electronic control	154
6.15	The Stanadyne distributor pumps	158
6.16	Stanadyne fuel delivery arrangements	164
6.17	The Stanadyne DS pump	166
6.18	The Bosch VE distributor type injection pump	169
6.19	Governing the Bosch VE type pump	173
6.20	Optional extras for the Bosch VE type pump	176

CHAPTER 7 COMBUSTION — 184

7.1	Thermodynamic characteristics of the diesel cycle	184
7.2	Differences between spark and compression ignition	187
7.3	Mixture preparation and ignition	188
7.4	Ignition delay	188
7.5	Cold starting	190
7.6	Cold starting aids	191
7.7	The three phases of normal combustion	193
7.8	Direct injection combustion chamber design	193
7.9	Injection viewed in detail	197
7.10	Injector holes and spray penetration	200
7.11	Mixing fuel and air	201
7.12	The generation of swirl	201
7.13	Indirect injection	204
7.14	Reducing heat loss from indirect injection combustion chambers	206
7.15	The small direct injection engine design problem	209
7.16	Direct injection and spark ignition compared	209
7.17	Direct injection, the ultimate aim?	211

CHAPTER 8	**EXHAUST EMISSIONS**	**212**
8.1	Reduction of emissions — conflicting requirements	213

Individual emissions in detail — 215
8.2	Oxides of nitrogen, NO_x	215
8.3	Unburnt hydrocarbons, HC	218
8.4	Carbon monoxide, CO	220
8.5	Particulates and black smoke	220
8.6	Particulate traps - general	224
8.7	Particulate traps in detail	226
8.8	Influence of fuel quality	230
8.9	Black smoke	230
8.10	White smoke	231
8.11	Emissions in practice	232

CHAPTER 9	**TEST CYCLES, SAMPLING AND ANALYSIS OF EXHAUST EMISSIONS**	**235**
9.1	Legislative control of diesel engine emissions worldwide	235
9.2	Units of measurement	235
9.3	Historical review	236
9.4	Test cycles	239
9.5	Sampling the exhaust gases	244
9.6	CO and CO_2	247
9.7	HC and NO_x	249
9.8	Measurement of particulates and black smoke	250
9.9	The Hartridge smoke meter	250
9.10	AVL 415 smoke meter	251
9.11	Pierburg smoke meter	253

CHAPTER 10	**OPTIMISING AIR INDUCTION — VARIABLE VALVE TIMING AND DIFFERENCES IN APPROACH FOR DIESEL AND GASOLINE ENGINES**	**255**
10.1	The Atkinson Cycle	256
10.2	History of variable valve timing	256
10.3	Two types of variable valve timing	257

Application to gasoline engines — 257
10.4	Early or late inlet valve closure	257
10.5	Late inlet valve closure (LIVC)	258
10.6	Early inlet valve closure (EIVC)	260
10.7	Problems associated with EIVC	264
10.8	Variable valve timing with control by throttle only	264

VVT mechanisms		265
10.9	Some simple systems	265
10.10	VPC, VLPC and VET systems	266
10.11	Mechadyne-Mitchell system	268
10.12	Control of the Mechadyne-Mitchell system	273

CHAPTER 11	**OPTIMISING AIR INDUCTION — INDUCTION PIPE TUNING**	**275**
11.1	The three effects	275
11.2	Resonant, or standing waves	276
11.3	Pipe end-effects	279
11.4	Frequencies, wavelengths and lengths of pipes	280
11.5	The Helmholz resonator	282
Engine induction system phenomena		284
11.6	Interference between pipes, or charge robbery	284
11.7	The inertia wave	285
11.8	Tuning the pipe to optimise the inertia wave effect	285
11.9	Tuning the pipe to optimise the standing wave effects	286
11.10	Harmonics of the standing waves	287
Practical applications		289
11.11	Telescopic induction systems	289
11.12	Vauxhall and Toyota two-stage induction pipes	290
11.13	Peugeot obviate inter-cylinder interference	292
11.14	The Honda variable volume induction system	294
11.15	Volvo siamesed pipe system	295
11.16	Some interesting alternatives	297

CHAPTER 12	**OPTIMISING AIR INDUCTION — TURBOCHARGING, AND THE COMPREX PRESSURE WAVE CHARGER**	**302**
12.1	Turbocharging	303
12.2	The turbocharger unit	305
12.3	Turbine nozzles and compressor diffusers	307
12.4	Potential for improving turbine efficiency	307
12.5	The compressor	309
12.6	Compressor surge	312
12.7	Compressor stall	313
12.8	Improving compressor efficiency	314
12.9	Valve timing	317
12.10	Turbocharger installation	317
12.11	Turbocharger characteristics and limitations	318
12.12	Turbocharger lag	319
12.13	Matching the turbocharger to the engine	320
12.14	The Comprex pulse charger	324

Index	329

Chapter 1

Crude oil derived diesel fuel

The chemistry of hydrocarbon fuels is dealt with adequately for our purposes in Vol. 1, so there is no need to repeat it here. On the other hand, the properties required for the hydrocarbons for use in diesel engines differ widely from those for gasoline power units. This difference arises primarily because the combustion processes are based on totally different principles. Fortuitously, diesel fuel inherently has mild lubrication properties such that it can be pumped by the injection equipment without causing seizure of the metal plungers in their barrels, in which they are hand lapped or machine lapped and selectively assembled to obtain an extremely close fit.

Whereas the gasoline engine draws in a ready mixed combustible charge of fuel and air which is ignited by a spark, the diesel unit is supplied with its fuel and air separately, and the two constituents then have to be mixed in the cylinder before being ignited by the heat generated during the compression stroke. Furthermore, while control over the gasoline engine is effected by throttling the flow of mixture into the cylinder, the diesel unit is controlled by increasing or decreasing the quantity of fuel injected per induction stroke into the cylinders, to regulate the overall rate of supply to the engine.

High thermal efficiency is the primary reason why almost all commercial vehicles in the medium to heavy range are powered by diesel engines. Despite their low output in terms of both power per litre and per kg, which translates into higher initial cost and greater bulk and weight for a given power output, diesel engines still offer lower overall cost of operation. This is especially so for local delivery trucks and vans, and for buses and other vehicles confined mainly to urban areas. Even long distance haulage vehicle operators benefit over the very large mileages that they operate, but this is largely, though by no means entirely, because the drivers are trained to use their gears to keep within the most economical speed range of the engines. Obviously, the heavier the loaded vehicle, the smaller is the proportion of its engine weight relative to the total. In the USA, where fuel costs are in any case much lower than in Europe, the diesel engine made much slower progress in replacing gasoline engines in the commercial vehicle field.

1.1 The diesel powered car

Although the first diesel vehicle on the roads of the UK was a Mercedes lorry in 1928, the first diesel car in series production was the 1936 Mercedes 260D. In the meantime, a one-off diesel powered Bentley special competed in the RAC Rally in 1932 at an average of 80 mph and 30 mpg.

In the past, the gasoline fuelled engine has been more attractive for cars because it has offered much better acceleration and top speed. This has been because not only is the specific power of the diesel engine inferior, but also it is heavier and more bulky, and therefore adds significantly to the overall weight of the vehicle. Other factors, which have been largely offset by improvements to diesel fuels and engines over the decade in question, include unpleasant smell, noise, problems in very cold weather and, in some areas, limited refuelling facilities for diesel vehicles. Moreover, the weight problem has been greatly ameliorated by the use of turbocharging, which can be applied to better advantage to diesel than gasoline engines.

By about 1990, owing to legislative pressures for reducing fuel consumption and threats of legislation to reduce the output of CO_2, the proportion of diesel powered cars was increasing. The rate of increase has been dependent to a major degree on relative costs of diesel and gasoline fuels. If figures for registrations of all European countries had been included in Fig. 1.1, it would

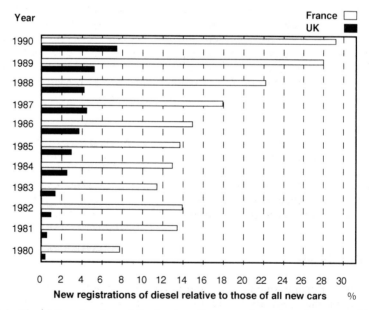

Fig. 1.1 The relative attractions of diesel and gasoline powered cars depends on price of fuel, in other words tax. If registrations in countries in which relative taxation levels have been changed during the ten year period had been included in the chart, some pronounced swings in demand would be apparent

be seen that, while registrations of diesel cars in countries, such as the UK, in which the taxation policy has been consistent over the period have had steady increased, those in which the cost advantage of diesel fuel relative to gasoline has markedly decreased show pronounced swings towards the latter.

Italy, for example, in 1985 increased its tax on diesel fuel from 31.5% to 67.3% of the price on the forecourt, so gasoline became significantly cheaper. This led to a drop in new registrations of diesel cars between 1984 and 1989 from a peak of over 24% to about 4.5%, though this was followed by a natural growth of just about 1% in 1990, as the earlier upward trend was then re-established. In 1992, diesel cars accounted for well over 12% of all registrations in the UK, and the increase in demand for diesel fuel for cars was about 22%, as against 7% for gasoline.

In 1977 in the USA, there were virtually no diesel powered cars but, with the steady rise in fuel cost and fears of a fuel shortage following the 1973 oil crisis, the market for diesel power expanded rapidly. By 1981, however, fuel prices were dropping again, and the demand for diesel cars fell steeply. In Japan, the demand has been increasing steadily since about 1976.

With the increasing application of turbocharging and other advances, specific power outputs have been improved, and engines operating over wider speed ranges and that are quieter have been developed. Furthermore, as production quantities have increased, prime costs have fallen. Exhaust smoke is often quoted as a disadvantage, but this can be virtually totally eliminated by rigorous control over the adjustment of the fuelling system. Improvements in both diesel fuels and additives have helped to overcome the problems of noise, smell and cold weather operation. Consequently, objections to diesel power for cars have been rendered irrelevant. Indeed, by the end of the decade 1980-1990, covered in Fig. 1.1, there were diesel cars on the road capable of 120 mph with accelerations comparable to those of gasoline powered cars. A good example is the 1991 Citroën XM powered by the 2.1 litre turbocharged diesel compared with the 2.0 litre carburettor and injection engines respectively. Their 0-100 km/h acceleration figures are respectively 11.1, 11.0 and 10.3 sec, and their top speeds 119, 120 and 127 mph.

From the environmental viewpoint, the diesel engine emits less than a third of the HC, about 1% of the CO and 30% less carbon dioxide than do gasoline engines. It has, however, a tendency to emit particulates and higher proportions of NO_x. For any given type of fuel, reduction in carbon dioxide output can be achieved only by increasing thermal efficiency. All matters related to emissions are covered in more detail in Chapter 8.

1.2 Diesel fuel

Whereas in Europe, one grade of diesel fuel is generally available for road vehicles, in the USA there are two, ASTM D1 and D2. Some of the requirements for fuels for diesel engines, such as high calorific value (energy content), are those needed also for gasoline power units, but most are entirely different. For customs purposes, the EEC definition of a diesel fuel is one

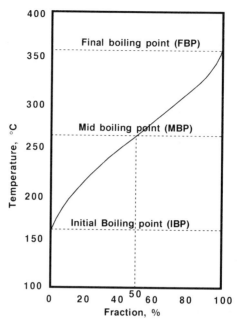

Fig. 1.2 The distillation curves for diesel fuels cover a range of fractions from about 150 to 360 deg C, as compared with about 30 to 210 deg C for gasoline

that contains a maximum of 65% distilled off at 250 deg C and a minimum of 85% distilled off at 350 deg C.

A typical analysis of a diesel fuel in the UK would show properties as follows:

Specific gravity	0.85	Cloud pt, °C	−5.5
Cold filter plugging pt, °C	−18	Cetane No.	51
Sulphur, %	0.22	Initial boiling pt, °C	180
Final boiling pt, °C	360	50% vaporisation, °C	280

Although distillation of crude oil can continue up to about 370 deg C, or slightly higher, before thermal cracking is liable to occur, diesel fuel mostly comprises fractions boiling off from approximately 250 to 355 deg C, Fig. 1.2, as compared with about 15 to 210 deg C for gasoline. These higher boiling point fractions contain about 20 times more sulphur than those from which gasoline is derived, so care has to be taken to remove it during refinement. Hydrocracking and catalytic cracking are used to convert fractions having even higher boiling points into hydrocarbons suitable for use as diesel fuels. Both hydrocracked and catalytically cracked fuels tend to have low cetane numbers (10 to 30). However, whereas the oxidation resistance in storage of hydrocracked fuels is high, that of catalytically cracked fuels is not, and therefore they tend to be slightly unstable.

1.3 Properties required for diesel fuels

The following are the properties that must be controlled when diesel fuels are blended:

Volatility	High volatility helps with cold starting and in obtaining complete combustion
Flash point	The lower the flash point the greater is the safety in handling and storage
Cetane number	This is a measure of ignitability. The higher the cetane number the more complete is the combustion and the cleaner exhaust
Viscosity	Low viscosity leads to good atomisation
Sulphur	Low sulphur content means low wear and a smaller particulate content in the exhaust
Density	The higher the density the greater is the energy content of the fuel
Waxing tendency	Wax precipitation can render cold starting difficult and subsequently stop the engine.

The properties of the fuel depend in the first instance on the source of the crude oil from which it is distilled. They vary as follows:

UK and Norwegian	Mainly paraffinic (Vol. 1, Section 1.1) and therefore of high cetane number. Calorific value relatively low and cloud point high. Sulphur content low to medium.
Middle East	Similar, but high sulphur content.
Nigeria	Naphthenic (Section 1.3, Vol.1). Low cetane number, cloud point and sulphur content. Calorific value medium.
Venezuela and Mexico	Naphthenic and aromatic (Section 1.4, Vol.1). Low cloud point and very low cetane number, but low to medium sulphur. Calorific value high.

Each of the properties in the lists above influences engine performance, so further explanation is called for.

1.4 Cetane number, cetane index and diesel index

Basically, *cetane number* is the percentage of cetane in a mixture of cetane and heptamethyl nonane (the latter sometimes referred to as alpha-methyl naphthalene) that has the same ignition delay, in terms of deg of rotation of the crankshaft, as the fuel under test. However, a more precise definition will be given later.

Ignition delay is important because, if it is too long, the bulk of the charge in the cylinder tends to fire simultaneously, causing violent combustion, Section 7.4. With a short delay, ignition is initiated at several points, and the flame subsequently spreads progressively throughout the charge. On the other hand, unless the injection is timed appropriately for the cetane number of the fuel to be used, rough running and other problems can still be experienced. Too high a cetane number can cause ignition before adequate mixing has occurred and thus increase emissions.

Cetane is straight-chain normal hexadecane ($C_{16}H_{34}$). On the other hand, heptamethyl nonane is a multiple branched alkane (combined with seven CH_3 radicals), and is defined as having a cetane number or 15. Middle East crude oils are a particularly good source for diesel fuel, because of the high proportion of alkanes and small proportion of aromatics that they contain. Alkanes are of better ignition quality than the aromatics.

The cetane number is defined precisely as the % n-cetane + 0.15 times the % of heptamethyl nonane contents of the blend of reference fuel having the same ignition quality as the fuel under test. Ignition quality is determined by varying the compression ratio to give the same ignition delay period for the test fuel and two blends of reference fuels. One blend has to be of better and the other of poorer ignition quality than the test fuel, but the difference between the two has to be no more than five cetane numbers. The cetane number is obtained by interpolation between the results obtained at the highest and lowest compression ratios.

Carrying out these laboratory engine tests, however, is not at all a convenient method of assessing the quality of a fuel, so two other criteria are widely used. One is the diesel index and the other the cetane index. The diesel index, which is obtained mathematically, is computed by multiplying the aniline point of the fuel by its by API gravity/100. Aniline point is the lowest temperature in deg F at which the fuel is completely miscible with an equal volume of freshly distilled aniline, which is phenylamine aminobenzene ($C_6H_5.NH_2$), a colourless oily liquid. API, stands for American Petroleum Institution, and deg API, measured with a hydrometer = (141.5/sp. gravity at 60 deg F) − 131.5. It is a measure of density for liquids lighter than water.

Cetane index is calculated from API gravity and its volatility, the latter originally taken as represented by its mid-volatility, or mid boiling point, T_{50} (50% recovery temperature). Since its introduction, the formula has been modified from time-to-time, to keep up with advancing fuel technology, and is now based on the density and volatility of four fractions of the fuel (those at the 10%, 50% and 90% distillation temperatures T_{10}, T_{50} and T_{90} respectively). According to ASTM D 4737 1988, the cetane index is given by:

$$CI = 45.2 + 0.0892(T_{10} - 215) + 0.131(T_{50} - 260) + 0.0523(T_{90} - 310) + 0.901B(T_{50} - 260) - 0.420B(T_{90} - 310) + 0.0049(T_{10} - 215)^2 - 0.0049(T_{90} - 310)^2 + 107B + 60B^2$$

where $B = e^{[-3.5(D-0.85)]} - 1$, and D is the density at 15 deg C.

Cetane index is usually better than diesel index as an indication of what the cetane number of a fuel would be if tested in a CFR engine in a laboratory, and it is certainly much less costly and time consuming to obtain. In general, alkanes have high, aromatics low, and naphthenes intermediate cetane and diesel indices.

A value of 50 or above for either diesel or cetane index is an indication that the combustion and ignition characteristics of the fuel are good. Values of 40 or less are totally unacceptable, and values below 45 undesirable. Low values mean that cold starting will be difficult, white smoke will be generated and the engine will be very noisy.

BS 2869: part 1: 1988 prescribes minimum limits of 48 for cetane number and 46 for cetane index. In Europe and Japan the minimum cetane number requirement is 45 and in the USA 40, the latter possibly being because a high proportion of their crude oil comes from Mexico and Venezuela. A reduction of from 50 to 40 cetane number leads to an increase in the ignition delay period of about 2 deg crankshaft angle in a direct and about half that angle in an indirect injection engine. Because of the higher energy contents of the lower cetane blends, the fuel consumption tends to become slightly lower as cetane number is reduced.

1.5 Tendency to deposit wax

In cold weather, a surprisingly small wax content, even as little as 2%, can crystallise out and partially gel a fuel. These crystals can block the fuel filters interposed between the tank and injection equipment on the engine, and ultimately cause it to stall. In very severe conditions, even the pipelines can become blocked and a thick layer of wax sink to the bottom of the fuel tank. Paraffins are the most likely constituents to deposit out as wax which, because they have high cetane numbers, is unfortunate.

The various measures of the tendency of a fuel to precipitate wax include *Cloud point*, which is is the temperature at which the wax, coming out of solution, first becomes visible as the fuel is cooled (ASTM test D 2500). Another test (ASTM 3117) is the *wax appearance point*, which is that at which the wax crystals become apparent in a swirling sample of fuel.

Then there is the *pour point*, which is the temperature at which the quantity of wax in the fuel is such as to cause it to gel (ASTM D 97). Checks on the condition of the fuel are made a 3 deg C intervals, by removing the test vessel from the cooling bath and tilting it to see if the fuel flows, and the *gel point* which is that at which the fuel will not flow out when the vessel is held horizontal for 5 sec. In practical terms, this translates roughly into a temperature 3 deg above that at which it becomes no longer possible to pour a fuel out of a test tube.

Other tests include the *Cold Filter Plugging Point* (*CFPP*) *of Distillate Fuels*, IP 309/80 and the CEN European Standard EN116:1981). This the lowest temperature at which 20 ml of the fuel will pass through a 45 μm fine wire mesh screen in less than 60 sec. In the USA, the *Low Temperature Fuel Test* (*LTFT*) is preferred to the CFPP test. These and other tests are described in detail in Section 2.4.

Although BS 2869: 1988 specifies a minimum cold filter plugging point (CFPP) of −15 deg C, it sets no low limit for cloud point. This seems odd because, when the filter becomes clogged it is too late to save the driver and vehicle from becoming stranded. Consequently, vehicle operators are calling for the inclusion of cloud point as a criterion for giving early warning that failures of vehicles in the field, perhaps far from base, could occur.

1.6 Density

Because it is related to energy content, density is significant. In units of kg/m^3, the densities of fuels obtained by the different refining processes are approximately as follows:

Straight run distilled	805 to 870
Hydrocracked gas oil	815 to 840
Thermally cracked gas oil	835 to 875
Catalytically cracked gas oil	930 to 965

Density, is measured by the use of a hydrometer with scales of specific gravity or gm/m^3. The sample should be tested at 15 deg C or the appropriate correction applied. Density is different from API Gravity, or deg API, in that the higher the number in deg API Gravity, the lighter is the fuel.

Because injection equipment meters the fuel on a volume basis, any variations in density will affect the power output. Moreover, at maximum power output, a high density fuel will produce more smoke as well as increase the power.

1.7 Volatility

The volatility of the fuel influences many other properties, including density, autoignition temperature, flash point, viscosity and cetane number. High volatility could cause vapour lock and lower the flash point, the latter having an adverse affect on safety in handling and storage. However, the higher the volatility the more easily does complete vaporisation of the fuel take place in the combustion chamber. Consequently, low volatility components may not burn completely and therefore could leave deposits and increase smoke. Within the range 350 to 400 deg C, however, the effects of low volatility on exhaust emissions is relatively small.

Surprisingly, the mid-range volatility has a marked influence on the tendency to smoke. This could be an outcome of the influence of this constituent of the fuel on injection and mixing, though cetane number can be influenced by the 50% distillate recovery temperature. In practice, it is the mix of volatilities that is most important: high volatility components at the lower end of the curve in Fig. 1.2 improve cold starting and warm up, while low volatility components at the upper end increase deposits, smoke and wear.

1.8 Viscosity

The unit of kinematic viscosity is the Stoke and is expressed in cm^2/sec, while that of absolute viscosity is the poise, which is the force required to move an area of 1 cm^2 at a speed of 1 cm/sec past two parallel surfaces that are separated by the fluid. For convenience the figures are usually expressed in centipoises and centistokes (cP and cSt). The two are related in that cP = cSt × density of the fluid. Viscosity of a fluid is determined by measuring the time taken for a certain volume of fuel at a prescribed temperature to flow under the influence of gravity through a capillary tube of a prescribed diameter. The SI units are m^2/sec, and the CGS or Stokes units are cm^2/sec.

Increasing viscosity reduces the injector spray cone angle and fuel distribution and penetration, while increasing the droplet size. It will therefore affect optimum injection timing. An upper limit has to be specified to ensure adequate fuel flow for cold starting. Lucas Diesel Systems quote a figure of 48 cSt at −20 deg C as the upper limit and, to guard against loss of power at high temperatures, 1.6 cSt at 70 deg C as the lower limit. BS 2869 calls for a maximum value of 5 cSt and a minimum of 2.5 cSt at 40 deg C. In Fig. 1.3, all these points are plotted and a viscosity tolerance band established.

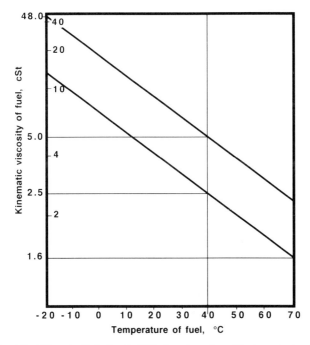

Fig. 1.3   Diesel Fuel Viscosity limits for the UK, plotted on a logarithmic vertical scale. The upper and lower limits on the 40 deg C line are those of BS 2869, while those at −20 and +70 deg C are recommendations by Lucas Diesel Systems. Fuels specified for winter operation would be biassed towards the lower limit and those for summer towards the higher limit

The viscosity of the average fuel lies about mid-way between the upper and lower limits.

Several consequences can arise from extremely high or low viscosity values. Too high a viscosity can cause excessive heat generation in the injection equipment, owing to viscous shear in the clearances between the pump plungers and their cylinders. On the other hand, if it is too low, the leakage through those clearances, especially at low speeds, can be excessive. Indeed, restarting following a brief shut down after operation at high load can become impossible. This is because, during the period when neither air is flowing over nor coolant flowing through the engine, the increase in temperature of the fuel locally due to conduction of heat to the injection system tends further to reduce the viscosity of the fuel in the pump.

1.9 Effects of different fuel constituents

There remain wide gaps in our knowledge of how the different constituents of fuels affect performance and emissions. However, white smoke positively increases as cetane number and volatility of the fuel are reduced. White smoke comprises mainly fuel but also some water droplets, and appears when the engine is started from cold and persists until the temperature has risen to the point at which the droplets are vaporised in the engine and remain so until well after they have issued from the exhaust tailpipe. The reason why the fuel droplets, although surrounded in the combustion chamber by excess oxygen, remain unburnt is that, in the cold environment, not only do they not evaporate but also their temperature never rises to that of auto-ignition.

Although cetane number increases with the density and volatility, it varies with composition of the fuel. Fuels having high cetane numbers are principally the paraffinic straight run distillates. However, because these have both high cloud points and low volatility, a compromise has to be struck between good ignition quality and suitability for cold weather operation.

One of the products of combustion can be black smoke. This is formed because the hydrogen molecules are oxidised preferentially so, if there is insufficient oxygen in their vicinity, it is the carbon atoms that remain unburnt. Aromatics have been said to produce black smoke, but the term aromatic is vague: basically they are structurally molecular rings. However, in diesel fuel, few so called aromatic molecules contain less than ten carbon molecules and, among these, most have only one aromatic nucleus. Among the di- and tri-aromatic compounds present, most have long tails and other molecular structures outside their rings, Fig. 1.4. Moreover, we do not know whether the aromatic nuclei burn independently, or if each nucleus and tail burns as a single unit.

High viscosity and density and low volatility also increase the tendency to produce black smoke. On the other hand, it has been suggested that this is misleading. Because the more volatile fuels have high API gravities (indicating low specific gravities), and their low viscosities allow more to

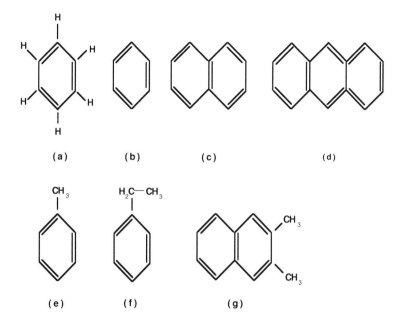

Fig. 1.4 Diagrams illustrating the characteristic ring shape molecular structures of aromatics and some of their more complex variants. The basic chemical formula of the mono-aromatic compounds is C_nH_n
(a) Formal representation of benzene (C_6H_6), which is a mono-aromatic compound
(b) An abbreviated, or shorthand, representation of benzene
(c) Naphthalene, $C_{10}H_8$, a di-aromatic compound (two rings)
(d) Anthracene, $C_{14}H_{10}$, a tri-aromatric compound (three rings)
(e) Toluene, $C_6H_5.CH_3$, obtained by substituting an alkane (formerly called paraffin) molecule for one of the hydrogen atoms in (a)
(f) Ethyl benzene, $C_6H_5.C_2H_5$, obtained by substituting a pair of alkane molecules for one of the hydrogen atoms in (a)
(g) 2.3 di-methyl naphthalene, $C_{10}H_6.(CH_3)_2$, with alkanes substituted for two of the hydrogen atoms

escape back past the injection pumping elements, the weight delivered to the cylinder per injection is smaller. For any given power, a certain weight of fuel must be injected, so a greater volume of the more volatile fuel and a longer injection period are therefore required and this might affect droplet size. Depending on the air movement and other conditions in the combustion chamber, such changes could have either a beneficial or adverse effect on combustion and therefore smoke generation. This could account for the conflicting reports on the subject of the relationship between fuel volatility and smoke generation.

Claims have been made that particulates come from fractions having high boiling points, but they has not been proved satisfactorily. This, however, introduces another problem for the refiner, which is that crude oils containing high proportions of the paraffins tend to have high sulphur contents and therefore adversely influence particulate content in the exhaust.

Even though more than 90% of the total mass of particulates in the exhaust is carbon, sulphur compounds are a problem and could become a significantly higher proportion of the total as advances in engine and injection equipment design lead to reduced carbon particulates from both the fuel and lubricating oil. Removal of sulphur is a costly process, and of course costly fuels mean higher prices for everything that is transported by road.

Changing the formulation of fuels can present unforeseen problems, and therefore cannot be put into effect without first carrying out extensive laboratory tests and field trials. Aromatics, for instance make a major contribution to the lubricity of the fuel so, their removal to reduce smoke can give rise to abnormally high rates of wear of injection pumps, especially in distributor type pumps in which all the work is done by only one or two plungers and perhaps a single delivery passage might be subject to severe erosion. Consideration has also to be given to the effects of fuel constituents on seals which, in injection systems, may operate at very high temperatures. Changes in fuel formulation for reducing exhaust emissions could cause nitrile rubber seals to shrink, harden and crack, and therefore have to be replaced by components made of a fluoro elastomer.

Chapter 2
Fuel quality and additives

Until the mid 1970s, the diesel fuel generally available in the UK and Europe was of a very high quality. Subsequently, on account of a shortage of appropriate crude oils, the quality worldwide showed a tendency to fall, though not as rapidly as had been expected. Since fuel reserves are being consumed at an increasing rate, the trend inevitably will remain downwards unless some economical way of synthesising high quality diesel fuel in the enormous quantities required can be developed. This trend has already led to the introduction of additives, many of which hitherto had been entirely unnecessary and therefore not even under serious consideration.

Additives represent not only added benefit but also added cost of fuels. Consequently, unless they are truly cost-effective, there is no incentive for the oil companies to use them. Some are sold in the aftermarket but, unless the purchaser knows the content of his fuel, he could be simply adding to what is already present at saturation level, and therefore the extra will make little or no difference. A possible exception is where a commercial vehicle operator has a large stock of fuel, bought in summer for bulk storage, and wishes to convert for winter use that which remains, by adding anti-wax additives.

For the oil refiners, extra processing and blending is a practicable, though not easy, alternative to additives for improving some of the properties of their fuels. Moreover it entails compromises. As indicated in Chapter 1, ignition quality can be improved by including more paraffins, but this detracts from the properties that are required for low temperature operation. Another possibility, for example, is increasing calorific value by adding aromatics, but these tend to cause smoke. The scope, therefore, is limited.

2.1 Matching additives to fuels

Even after engines have been optimised as regards emissions and fuel economy, increasingly stringent regulations are unlikely to be met unless fuel quality is maintained. Because of the many interacting variables involved, blending base stocks to obtain high quality in a fuel is a complex operation. It can, however, be facilitated by the introduction of additives of types and in quantities dependent upon the blend to which they are applied.

A complicating factor as regards using additives effectively is that crude oils from different parts of the world have different characteristics as set out in Section 1.3. At the same time, the blends demanded vary from country-to-country. For example, in the USA a high proportion of the heavy fractions from the crude oil are taken for domestic and industrial heating. However, this is changing, as also are the relative quantities of gasoline, gas turbine and diesel fuel required in the different countries. Climate, too, is an important consideration, especially as regards tendency to wax precipitation. To meet all these demands, the oil companies take crudes from different sources, blend the distillates to make up fuels of various types and process them differentially to match the requirements of each market. Yet another problem is that the fuel characteristics needed vary with engine design.

Incidentally, the units used to express quantities of additives in fuels are generally parts per million (ppm). The slightly larger quantities may be expressed as percentages, so 0.1%m or 0.1%v are 1000 parts per million by mass or by volume respectively.

2.2 Additives in premium grade fuels

One of the early entrants (in 1988) into the multi-additive diesel fuels market was Shell's Advanced Diesel. This contains several additives, one of which has raised its cetane number from the 48 required by British Standards, and the minimum of 50 for Shell's base fuel, to typically between 54 and 56.

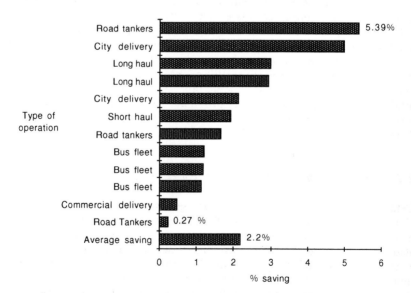

Fig. 2.1 Results of tests on fuel consumptions in a number of fleets on different types of operation. Dependent as they are on variables including driver habits, the individual results show a considerable degree of scatter

FUEL QUALITY AND ADDITIVES

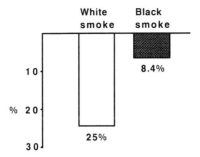

Fig. 2.2 Smoke emissions tests have shown that, in comparison with a commercially available fuel, termed the base fuel, Shell Advanced Diesel reduces white smoke on average by 25% and black smoke by 8.4%

Among the others are a corrosion inhibitor, anti-foam, cold flow and re-odorant additives. The benefits claimed include lower noise, 3% better fuel economy, Fig. 2.1, 8.4% less black and white smoke, Fig. 2.2, a general improvement in overall engine performance and durability, and a reduction in downtime.

The 3% better fuel economy is based on carefully monitored trials over a period of 7 months with 24 Cummins powered Ford trucks on heavy duty long haul operations, by Overland Express, in Canada. Of these vehicles 12 were run on Shell Advanced Diesel, 12 on one commercially available fuel and 12 on another. The average distance travelled per truck was 85,000 miles and the total fuel consumed was 810,000 litres.

Of all the additives available, the most obviously important to the operator are the cetane improvers and those that help to overcome the tendency to wax precipitation in winter. Others used include anti-oxidants, combustion improvers, cold flow improvers, corrosion inhibitors, detergents, re-odorants and anti-foamants. Stabilisers, dehazers, metal deactivators, corrosion inhibitors, biocides, anti-icers, demulsifiers. Anti-static additives are used too, but mainly to benefit the blenders by facilitating storage, handling and distribution.

2.3 Cetane No. and combustion improvers

Cetane number is a measure of the ignitability of the fuel, Fig. 2.3. Consequently, a low value may render starting in cold weather difficult and increase the tendency towards the generation of white smoke, Fig. 2.4. It also increases the ignition delay (interval between injection and ignition). As a result, because fuel is injected into the combustion chamber for a longer time prior to ignition, the ultimate rate of pressure rise is more rapid, and the engine therefore noisy. Also, because there is less time for the fuel to burn before the exhaust valve opens, the hydrocarbon emissions will be increased.

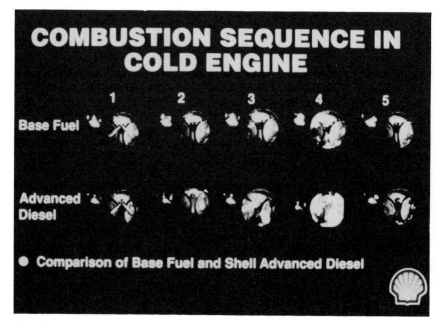

Fig. 2.3 Combustion sequences in the cylinder of an engine running on, top, a commercially available alternative fuel and, bottom, Shell Advanced Diesel. These pictures were taken through a quartz window in the piston, so the two part circles within the boundaries of the combustion chamber are the inlet and exhaust valve heads. The bright spot outside, on the left, of each combustion chamber picture is an illuminated pointer above degree markings on the flywheel, which are too small to be indentifed visually here

On the other hand, if the cetane number is higher than that for which the injection system is timed, power will be lost because a high proportion of the pressure rise will occur when the piston is at or near tdc. Furthermore, the fuel might ignite before it has mixed adequately with the air so, again, smoke and hydrocarbon emissions may be high. Consequently, fuels having high cetane numbers perform best when the injection is retarded. On the other hand, with too much retard, there will not be enough time for complete combustion, and smoke and HC emissions will be the outcome.

Since high cetane numbers are difficult to achieve, the regulations in most countries specify only low limits. This is because the supply of diesel fuel obtained by straight distillation is generally inadequate to satisfy demands, especially in the USA where the demand for gasoline is heavy, and the diesel fuel therefore has to be augmented with cracked products which, characteristically, have low cetane values. Indeed, while the minimum cetane number specified in Europe is mostly 45, in parts of the USA at the time of writing it is only 40. However, in California, there are already calls for raising the limit to 60.

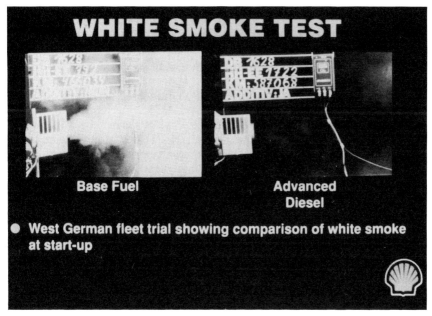

Fig. 2.4 Photograph taken during a West German fleet test, to compare the white smoke emission from an engine running on, left, an alternative commercially available fuel and, right, Shell Advanced Diesel. White smoke, a mixture of steam and unburned hydrocarbons, is reduced by increasing cetane number and, at low cetane numbers only, by increasing volatility

Additives used as cetane improvers include mainly alkyl nitrates. Even though it has been suggested that some nitro compounds might be toxic, this has not yet been proved satisfactorily. Other substances that decompose easily and form free radicals at high the temperatures in combustion chambers can be employed. These include ether nitrates, nitroso compounds and some peroxides. With *iso*-octyl nitrate, which is the most commonly used, it is possible to obtain cetane number improvements of between 2 and 5 numbers, depending on the base fuel and the quantities of additive employed. Nitrates of *iso*-propyl, amyl, octyl or hexyl are also used. Unfortunately, it is the fuels that have the lowest cetane numbers that tend to respond least to cetane improvers.

Combustion improvers differ from cetane improvers in that they are mainly organic compounds of metals such as barium, calcium, manganese or iron and are catalytic in action. Barium compounds could be toxic, so interest now centres mainly on manganese and copper compounds. Although they produce metal-based particulates, they also lower the autoignition temperature of the carbon-based deposits in both the cylinders and particle traps, causing them to be more easily ignited.

2.4 The wax problem

All additives for cold weather operation modify the shapes of the wax crystals, which otherwise are flat platelets tending to gel together. Paraffins are most likely to deposit out as wax. This is unfortunate because they have high cetane numbers.

As an indicator, cloud point has the disadvantage of dependence on the tester's judgment. In another test (ASTM 31117) a mechanical apparatus induces a swirl in the fuel sample, and the wax appearance point is that at which the wax crystals render the swirl visible. This has the advantage of eliminating any temperature gradient that otherwise might be present in the fuel, so the wax crystals are uniformly distributed and at least the temperature reading can be more reliable.

While most fuels are unusable well above their pour points, many can still be used at temperatures even well below their cloud points. Consequently, the Cold Filter Plugging Point (CFPP) of Distillate Fuels, IP 309/80) has been adopted in many countries, and has been the basis of the CEN European Standard EN116:1981.

Different countries specify cold filter plugging points ranging from about $-5°C$ in the Mediterranean region to $-32°C$ in the far north. In practical terms, the basic consideration is that the engine shall start at the lowest overnight soaking temperature likely to be experienced in service. Once it has started, if the filter becomes partially blocked, the rate of flow could be such that the return flow of warm fuel to the tank is reduced, and therefore so also would be the rate at which the temperature of the fuel in the tank is raised. Naturally, if the engine warms up rapidly, the wax in the fuel system will tend to melt and dissolve sooner.

Since paper element filters are now widely used, the relevance of wire mesh filters is open to question. Consequently, the CEN is debating the desirability of introducing a test called the Simulated Filter Plugging Point (SFPP). Moreover, the CFPP does not correlate with vehicle performance in the USA, where fuel blends and quality, modes of operation, and fuel system components and temperatures are different. Consequently, the Americans have developed the Low Temperature Fuel Test (LTFT). It differs in detail from the European test in that it requires 200 ml of fuel to be cooled and drawn by a depression of 6 in of mercury through a 17 μm screen. The LTFT is the temperature at which 180 ml of fuel passes through the screen in less than 60 sec.

Another way of estimating operational performance of a fuel is to combine the cloud point and the difference between it and the pour point, to obtain a Wax Precipitation Index (*WPI*). The formula is as follows:

$$WPI = CP - 1.3(CP - PP - 1.1)^{0.5}$$

2.5 Cold weather additives

Cold flow improvers, or wax anti-settling additives (WASA), were among the first additives to have been of practical value in diesel fuels. However, they are difficult to distribute uniformly throughout the fuel in quantities adequate for them to be effective. This has led to the development of new additives for modifying the shape of the wax crystals to enable them to pass the CFPP test. These additives, which include ethylene vinyl acetate, polyolefin ester, and polyamide, are also the basis for what some aftermarket producers call flow improvers.

They modify the shapes of the wax crystals, which otherwise are flat platelets tending to gel together. There are three types of modifier: pour point depressants, flow improvers and cloud point depressants. Which of these is used depends basically on local requirements and the type of wax to be treated. The latter is mainly a function of the boiling range of the distillates and the country of origin of crude oil.

As indicated in Section 1.2, cracked products contain high proportions of aromatics, which have low cetane numbers, but contain less wax and, moreover, dissolve what wax there is more readily than straight distilled fuels. Fuels with a narrow boiling range form large wax crystals that are less susceptible to treatment by additives than the smaller and more regular shape crystals formed in fuels having a wider boiling range.

From Stoke's Law we deduce that the rate of settling of crystals is directly proportional to the square of their diameter times the difference between their density and that of the fluid, and is inversely proportional to the viscosity of the fluid. Consequently, small crystal size is the overriding need.

Pour point depressants for diesel fuels were introduced in the 1950s. They interact with the wax crystals to reduce their size and modify their shape: some multi-axial needle crystals are introduced between the platelets, which also become thicker, so it is more difficult for them to interlock.

More effective are the flow improvers. These cause small multi-axial needle crystals to form, instead of larger platelets. Moreover, by virtue of the presence of some of the additive molecules between the crystals, they tend not to adhere to each other. Although they will pass through wire gauze strainers, they are stopped by the finer filters used to protect the injection pumps and injectors. However, by virtue of their small size and the multi-axial arrangement of the needles, the unaffected liquid fuel can still pass between them. Flow improvers currently in use include olefin-ester copolymers such as ashless copolymers of ethylene and vinyl acetate.

The small compact wax crystals tend to settle in the bottoms of tanks. This is more of a problem in storage than vehicles, but wax anti-settling additives (WASA) can nevertheless play a useful part in the avoidance of wax enrichment as vehicle fuel tanks become empty, especially in very cold climates. These additives are usually employed in proportions ranging from 100 to 500 parts per million.

Their polymeric active ingredients, are similar to those of the flow improvers. They modify the crystal formation, by both forming nuclei and arresting growth. As the temperature of the fuel falls through the cloud point, small wax crystals grow on the nuclei and, subsequently, other additive molecules attach to their surfaces and block further growth. The outcome is small crystals. Primarily, these additives improve cold filterability, but they also lower the pour point.

Cloud point is an indicator of the quality of the base fuel. Although a cloud point depressant by itself lowers the cold filter plugging point of a base fuel, using it in a fuel containing also a flow improver may have the opposite effect. The improvement obtainable by olefin-ester copolymer cloud point depressants is generally only small, of the order of about 3 deg C and they are costly, so they are unattractive, except where cloud point is included as part of a diesel specification and the blender therefore wishes to lower it.

2.6 Dispersants and corrosion inhibitors

Dispersant additives include ashless polymers and organic amines. Their function is to restrict the size of the particles formed within the fuel and, additionally, to remove them from the surface. If not used continuously, however, they may dislodge gums which can block filters. Then there are dispersant modifiers, or detergents, such as polyamides, polyisobutane and succinimides. Their function is to keep the surfaces of the combustion chambers and injection nozzles clean but, if used to excess, can actually cause gums to form.

In a different category, corrosion inhibitors are precisely what their name implies. These serve not only to protect fuel system components but also bulk storage tanks and barrels. Alkyl phosphate is perhaps among the most commonly used inhibitors.

The fact that, like metals for example, fuel surfaces can be oxidised in contact with air is widely overlooked. Such oxidation can cause the formation of gums, sludges and sediments, but it can be reduced or prevented by adding surface-active sulphonates and polymers or alkylated phenol. These, however, must be added immediately after refining the fuel and while it is still warm.

2.7 Detergents and anti-corrosion additives

Detergents are used mainly to remove carbonaceous and gummy deposits from the fuel injection system. Gum can cause sticking of injector needles, while lacquer and carbon deposited on the needles can restrict the flow of fuel, distort the spray and even totally block one or more of the holes in a multi-hole injector, Fig. 2.5. The outcome can be misfiring, loss of power, increases in noise, fuel consumption, HC, CO, smoke and particulate emissions in general, Fig. 2.6. Furthermore, starting may become difficult, because the fuel droplets have become too large owing to the reduction in flow rate.

The detergents used are mainly amines, amides and imidazones, at rates of about 100 to 200 ppm. Sometimes included with them, to help dispersal of the particulates, are about 200 ppm of alkenyl succinimide, or about 600 ppm of hydrocarbyl amine polymeric dispersants.

Detergent molecules are characterised by, at one end, a head comprising a polar group and, at the other, an oleofilic tail. The arms of the polar group

FUEL QUALITY AND ADDITIVES

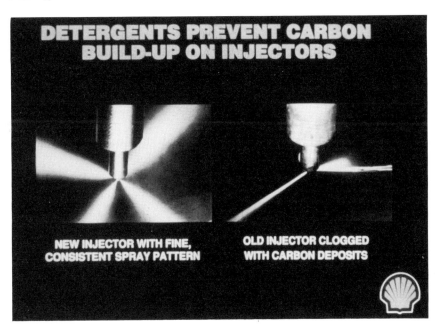

Fig. 2.5 Effects of carbon deposits on diesel injector sprays: left clean; right sooty

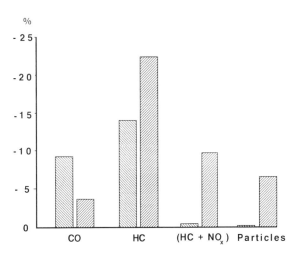

Fig. 2.6 Reductions of exhaust emissions during road tests on diesel powered cars run on Shell Advanced Diesel as compared with the alternative fuel. The measurements were made running to the ECE.15 procedure on a chassis dynamometer immediately after each of the vehicles had competed its mileages on the road. In each case, the left hand column indicates percentage clean-up by Shell additive after 4,000 km while the right hand column represents percentage of actual deposits measured after 15,000 km

FUEL QUALITY AND ADDITIVES

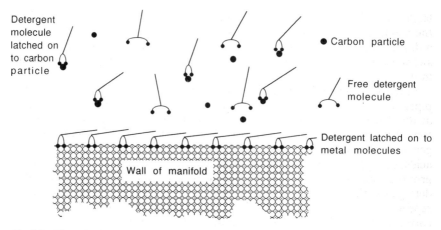

Fig. 2.7 The polar heads of the detergent moleculs latch on to the metal surfaces and particulate molecules. On the former, they form a barrier film preventing further deposition and, incidentally, affording a degree of protection against corrosion, while those that latch on to the particles are swept away with the fuel because their oleofinic tails carry them into solution in it

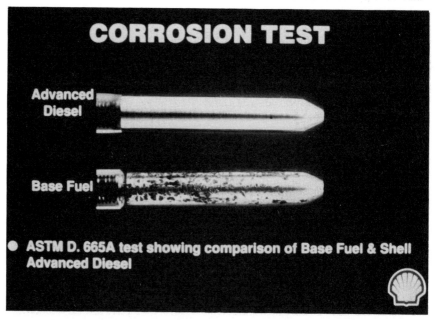

Fig. 2.8 Effects of corrosion on specimens tested to ASTM D.665A in, top, Shell Advanced Diesel and, bottom, the commercially available alternative fuel

FUEL QUALITY AND ADDITIVES

latch on to the metal and particulate molecules, Fig 2.7. Those attached to the metal form barrier films inhibiting deposition and, incidentally, offering a degree of protection against corrosion, while those on the particulates are swept away with the fuel because the oleofilic tails carry them into solution in it.

Anti-corrosion additives (perhaps about 5 ppm) are mainly used to protect pipelines in which diesel fuel is transported, but only trace proportions remain at the delivery end. Therefore, if vehicle fuel system protection is required too, the treatment must be heavier. The actual chemicals employed are generally esters or amine salts of alkenyl succinic acids, alkyl phosphoric acids, or aryl sulphonic acids. These additives are surfactants having a polar group at one end and an oleophilic/hydrophobic tail. As in the case of the detergents, the polar heads attach themselves to the metal, but the water repellent tails form an oily coating over the surface to protect it against corrosive attack. In Fig. 2.8, test samples that have been left over a long period in the base fuel are compared with those similarly left in Shell Advanced Diesel.

2.8 Anti-foamants and re-odorants

To allow the fuel tank to be completely filled more rapidly, and to avoid splashing of diesel fuel on the clothes of the person filling the tank, anti-foamants can be used, Fig. 2.9. They are usually silicon surfactant additives applied in quantities of between about 10 and 20 ppm.

Fig. 2.9 Comparison between the foaming of, left, the alternative fuel and, right, Shell Advanced Diesel when delivered through a dispenser nozzle into a glass container

A factor influencing the introduction of re-odorants has been the increasing numbers of diesel cars coming into use. Unlike truck drivers, car owners are unaccustomed to the lingering and unpleasant smell of diesel fuel. The problem is less serious with gasoline, mainly because of its rapid rate of evaporation. Elimination of the smell is undesirable because it facilitates the detection of leaks, so the aim is at modifying it by partially masking it with a more acceptable odour. Again, the rate of treatment is usually about 10 to 20 ppm.

Chapter 3
Alternative fuels

As mentioned in Vol. 1, Section 3.4, diesel fuel has been produced from coal. Potential substitutes envisaged for diesel fuel include vegetable oils such as rape seed methyl ester (RME), and fuels prepared from sunflower, safflower, babacu, corn, cotton, peanut, soya and castor oils. They have been particularly attractive as alternatives when crude oil has been in short supply, such as in World War 2, and more recently to enable farmers in developing countries to be self sufficient as regards supplies for their tractors and to avoid the need to import either crude or refined oils. Most are currently either actually in use in some parts of the world or the subject of research and development. They are economically attractive, however, principally in countries having no natural reserves of crude oil or gas, and whose climate is so favourable for agriculture that several crops can be harvested per growing season.

Viscosities of vegetable oil derived diesel fuels are higher and their energy contents about 10% lower than that of their crude oil based equivalents, but their auto-ignition properties are much the same. Problems commonly experienced are carbonaceous and gummy deposits in combustion chambers, lower smoke limits and filter blockage in cold weather. Incidentally, a recent discovery by the Scottish Agricultural College is that, by adding alcohol to rapeseed oil, glycerol is made to separate out, thus reducing the tendency to gum deposits in the engine.

With vegetable oils, emissions of HC and NO_x could be higher too. However, these might be overcome by injectors designed specifically for the fuel or the use of anti-oxidant, detergent and other additives. To economise in the consumption of products derived from crude oil, a 20% blend of vegetable oil with diesel fuel could be practicable. Blending in this way tends to reduce the emissions and smoke levels, though power output will still tend to be lower and separation of the blended constituents could occur in cold weather. Again, ways might be found of avoiding them.

Indeed, most of the problems no doubt could be overcome with further development. For example, additives might be used to combat deposit and gum formation and cold weather, as well as the problems already mentioned. Moreover, so far only oils already commercially available for other purposes have been used, so it is not beyond the bounds of possibility that some more suitable vegetable, currently growing unnoticed yet plentifully in the wild, might be discovered.

A point that should not be overlooked when alternative fuels are under consideration is their lubrication properties. The lubricity of alcohols in particular is poor. Lubricity is of vital importance since injection pumps are lubricated by the fuel. It is not impossible either to put lubrication oil in the sump of an in-line pump or to draw upon the engine pressure lubrication system. With the lattter arangement, however, it is impossible or, at best extremely difficult, to avoid fuel contaminating the oil in the engine, especially if the system is utilised to lubricate the pump plungers.

3.1 Alcohol fuels and blends for diesel engines

In some countries, waste products from food crops can be converted into alcohol or, alternatively, special crops can be grown solely for this purpose. Work has been done, by Volvo, Scania and others, on the use of methanol and ethanol as diesel fuels. Volvo installed experimentally two injectors, one for ethanol and the other for the hydrocarbon fuel, in each cylinder of one of their engines. It had to be started on diesel fuel because of the low cetane number (poor ignitability) of ethanol. However, the cooler combustion obtained during operation with the ethanol injector reduced both soot and NO_x emissions.

As the load increased, diesel fuel was injected in increasing proportions until, at maximum power output, it amounted to 90% of the total. Under typical running conditions, however, about 75–80% of the energy required to propel the truck was obtained from the ethanol. The penalties were two tanks, that for ethanol having a larger overall capacity, owing to the low calorific value of this fuel relative to that of diesel fuel, and white smoke emission from the exhaust when the engine was cold. Currently, dual injection would be too costly for commercial applications. On the other hand, shortage of diesel fuel of adequate quality might change the economic parameters to such an extent as ultimately to render it attractive. The alternative of blending the two fuels would probably exacerbate cold starting difficulties.

Alcohol has a low Cetane number, which could be offset by blending it with diesel oil. The autoignition temperatures of the alcohols, at about 1000 deg K, are almost double those of diesel fuels. Ignition improvers can be used, but in proportions as high as 15–16%. For starting, even with the engine at ambient temperatures, high compression ratios and glow plugs would be required. As regards the injection equipment, lubrication of the plungers might present problems and, because of the high volatility of the fuel, so could vapour lock and cavitation.

Even when blended with diesel oil, the alcohols have the disadvantage that they tend to absorb moisture, in which they dissolve and then separate out in the bottom of the tank. Consequently, unless measures are taken to protect the metal components in the fuel system, trouble can be experienced with corrosion. The air:fuel ratio required for their complete combustion is significantly lower than that for the hydrocarbon fuels, so the injection system has to be calibrated appropriately if combustion problems are to be avoided.

ALTERNATIVE FUELS

On the other hand, because of its lower energy content, if the power output is to remain the same as with conventional diesel fuels, larger quantities must be injected per cycle.

Scania, in the belief that currently no known fuels other than alcohols offer any long term potential as an alternative to diesel oil, ran practical tests with ethanol. They were undertaken in Brazil, where it is relatively easily produced in quantity from forest and agricultural crops. The conclusion was that, even there, the cost of ethanol is too high for it to be acceptable for general use. Scania confirmed that, partly because the latent heat of vaporisation of both the ethyl and methyl alcohols is so high, alcohols are difficult to ignite. Of the suitable ignition improver additives, nitrates are believed not to be a health hazard, but this has been difficult to prove, so polyethylene glycol might be the preferable alternative.

Scania found the NO_x and particulate output with alcohol to be lower than with diesel fuels. However, aldehydes, the effects of which are not yet fully understood, were more prominent in the exhaust gases. On the other hand, alcohol, has the advantage that the CO_2 produced during its combustion is taken back from the atmosphere for the production of more alcohol, on a closed cycle basis.

3.2 Prospects for methanol

The energy contents of various alternative fuels are clearly illustrated in Fig. 3.1, reproduced from the Scania publication 'The Diesel Engine and The Environment'. Methanol offers reductions of all emissions, and can be produced in quantity from either natural gas or coal at a price per litre ranging from 16–60% higher than that of the conventional liquid

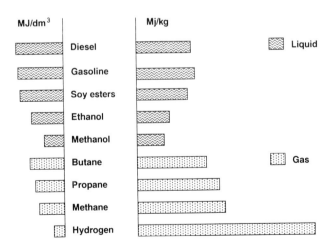

Fig. 3.1 A diagrammatic representation of the energy contents of various alternative fuels

Table 3.1 SPECIFIC GRAVITY AND DENSITY IN LB/GAL OF VARIOUS FUELS

Methane 0.30/2.50	Ethane 0.374/3.11	Propane 0.508/4.23	Butane 0.584/4.86
Pentane 0.631/5.25	Hexane 0.664/5.53	Heptane 0.668/5.73	Octane 0.707/5.89
Nonane 0.722/6.01	Decane 0.734/6.11	Undecane 0.744/6.19	Dodecane 0.753/6.27

hydrocarbons. However, if it were to be substituted entirely for diesel oil, some difficulty would be experienced in producing it in quantities adequate to meet demands. Moreover, from the investigations conducted by Volvo, their Chief Engineer, Per-Sune Berg, has concluded that the cost of methanol expressed in terms of $ per tonne-km would probably be double that of diesel fuel, the tank volume required $2\frac{1}{2}$ times and its weight double as much (the specific gravities of a range of fuels are given in Table 3.1). On the other hand, the energy efficiency of producing methanol from natural gas is 60–65%, and from coal 47–52%, which is higher than that of conversion of coal, by the Fischer Tropsch process, into either gasoline or diesel fuel. A good summary of the merits and disadvantages of methanol as a fuel has been given by Kowalewicz, *Proc I.Mech.E, Vol 207, D1*, 1993.

3.3 Driveability

As regards diesel engines, a formula that is used by Perkins Engines for obtaining a figure for comparing the driveability of vehicles with power units installed with appropriately matched transmissions is as follows:

$$\text{Max. torque} \times (\text{revs at peak power} - \text{revs at peak torque})$$

In other words, driveability is a function of both maximum torque and the engine speed range between maximum power and maximum torque. However, it must be appreciated that, for a truck driver, the driveability of his vehicle is to a major extent a function of how many gear shifts he has to make during his journeys. Ideally, therefore, he requires an engine the torque of which rises markedly as his speed falls from that at which maximum power is developed to that for maximum torque.

3.4 Gaseous alternatives

Natural gas is available in large quantities only in some parts of the world. It would considerably reduce particulates and is superior as regards the other emissions, but not so markedly. Its use would entail a change to spark ignition, fuel tanks would have to be 6 times as big and the weight of the fuel between 3.5 and 7 times as great.

Although not currently regarded as an alternative fuel for compression ignition engines, it is widely used in some countries in spark ignition engines, instead of diesel engines, for powering both light and, more especially, heavy commercial vehicles operated on local services for both passengers and delivery of goods. Indeed, it could become increasingly attractive as more stringent regulations are introduced to enforce a reduction in output of CO_2. Moreover, some research work is being carried out on natural gas injection, so compression ignition is perhaps not entirely ruled out. An analysis in detail of this alternative fuel is therefore appropriate.

3.5 LPG and LNG

Hitherto, as compared with liquefied petroleum gas (LPG), liquefied natural gas (LNG) has not been widely utilised in the UK, despite its being distributed in gaseous form to both domestic and industrial users throughout the country. One reason for the lack of interest has perhaps been that many still confuse what used to be called town gas, produced from coal, with natural gas, which is piped at 1000 lbf/in^2 from the North Sea fields to British Gas. Natural gas is in general over 90% methane, as compared with, on average, 48% H_2, 8% CO, 30% methane plus some N_2 and olefins of town gas.

Another reason is that LPG, mainly propane with some butane and, in very small quantities, other light hydrocarbons, can be stored in liquid form at atmospheric temperatures and therefore is not difficult to contain on a vehicle. LNG, on the other hand, cannot. Indeed, hitherto it has been virtually ruled out as an alternative fuel, because it cannot be liquefied at normal ambient temperatures. Consequently, it would have to be contained at low temperatures in thermally insulated vessels designed to withstand very high pressures, so questions of safety arise, not only in the event of accidents, but also in storage and distribution.

3.6 ANG, low pressure containment of natural gas

Now however, this gaseous fuel can be contained in a relatively dense form at atmospheric temperature and at a relatively low pressure. In this form it is termed adsorbed natural gas (ANG), because it is adsorbed in activated carbon. This method of containment has been developed to the point of commercial practicability as a result of a great deal of research and development of gas adsorption systems by British Gas at its London Research Centre, jointly with the Department of Chemistry and Engineering of the Royal Military College of Canada. Details and practical implications of this joint research are covered in a chapter entitled 'Natural Gas Adsorbed on Carbon', by N.D. Parkyns and D.F. Quinn, in a book entitled 'Porosity in Carbons – Characteristics and Applications', Edward Arnold. Vehicle installation development has been done by the British Gas Midlands Research Centre.

While LPG is stored in the fuel tank at a pressure of less than 7 bar and evaporates at about 0.5 bar, ANG calls for the rather higher pressure of 35 bar, approximately 500 lbf/in^2. Even so, this is still remarkably low since

LNG boils at approximately 111.7 deg K and its critical temperature, above which it cannot be liquefied no matter how high the pressure, is about 190.5 K.

3.7 Compressed and adsorbed natural gas (CNG and ANG)

Natural gas which, as previously mentioned, is mostly at least 90% methane, Table 3.2, should not be confused with biogas, which is obtained by fermentation of sewage and rubbish in general. The latter contains high proportions of impurities, including CO_2. Because of the low boiling point and critical temperature of natural gas, and also its low energy density per unit of volume (3.22 MJ/m^3 compared with about 42.7 and 43.7 MJ/kg for gasoline and DERV), it is difficult to contain in road vehicles in the quantities needed for their fuelling. Composition and energy densities of some fuels used for road transport are compared in Tables 3.3 and 3.4.

Of interest is the high energy density of LNG relative to that of methanol, which has been widely mooted as an alternative fuel for diesel engines. In a nutshell, it can be said that, in terms of energy content, 1 kg of natural gas is the equivalent of 1.33 litres of gasoline and 1.22 litres of diesel fuel. At

Table 3.2 COMPOSITIONS OF NATURAL GAS FROM DIFFERENT SOURCES

Component % by volume	Leman Bank, UK, North Sea	Source Frigg Norway	Groningen, Netherlands
Methane	94.81	95.72	81.5
Ethane	3.00	3.37	2.90
Propane	0.55	0.03	0.40
Butane	0.19	0.01	0.15
$>C_4$	0.14	0.02	0.10
Carbon dioxide	0.04	0.31	1.00
Nitrogen	1.22	0.53	14.00

Note: the content of hydrocarbons $>C_4$ can be regulated before transmission to customer, the sulphur level is approximately 75 to 180 ppm, and pipeline gas is dried to a water content of 75 to 180 ppm by volume.

Table 3.3 ENERGY DENSITIES (MJ/LITRE) OF GASEOUS FUELS AND BATTERIES

Natural gas			ANG (35 bar)		Battery
20 bar	100 bar	200 bar	Current	Future	
0.8	3.8	7.6	3.8	~6	0.2–0.3

Table 3.4 ENERGY DENSITIES (MJ/LITRE) OF LIQUID FUELS

Diesel	Gasoline	Methanol	LNG	LPG
37	32	16	23	29

first sight, this may appear to have been expressed in an odd mixture of units, but natural gas can be anything from a liquid to a gas at almost any density, according to its pressure and temperature, hence the need to quote its energy density in terms of its weight. The relative densities of the gasoline and diesel fuels is of the order of 0.74 and 0.85 respectively, or 0.98 and 1.0 kg per litre though, in practice, the figures in the latter units are hardly meaningful. Perhaps the following are more readily comprehended comparisons: gasoline has an energy content of about 32 MJ/litre, while that for CNG is approximately 8.8 MJ/litre and for LNG 23 MJ/litre.

For use as a fuel for road vehicles, natural gas hitherto has had to be compressed to about 20 MPa (3440 lb/in^2 or 200 bar), which increases its density to 140 g/litre. However, the pressure vessels in which it is contained are both heavy and costly. Even so, there are today about 700,000 natural gas vehicles, mainly in commercial operation, and despite the fact that they have run about 430 million miles, there has been no experience of burns, other injuries or fatalities arising directly out of the use of this gas as a fuel. On this basis, safety is claimed to be one of the assets of natural gas as a fuel.

Even greater safety is assured by the previously mentioned recent development, which is storage of this gas at 35 bar. This is done by adsorbing its molecules into, and thus containing them in, micropores in the surface of carbon. Dr N.D. Parkyns, of the British Gas London Research station says that it should be feasible to adsorb approximately 150 volumes of gas at 35 bar in 1 volume of activated carbon. This is equivalent to the same volume of compressed natural gas but at 150 bar, though this very high level of adsorption has not yet been achieved on a commercial basis. The range of operation of the vehicle would be about 75% of that which would be obtained if the compressed gas and perhaps a sixth of that if diesel fuel were used, but at a significantly lower cost. However, bearing in mind the sometimes violently fluctuating costs of crude oil, the relative costs of operation, though admittedly dependent on levels of taxation, could at any time change dramatically in favour of natural gas.

3.8 Carbon, its potential and properties

When carbon is produced in a vacuum or inert atmosphere, it has micropores totalling a very large proportion of its volume. Although many other adsorbent materials have been investigated, none has so far been found to be better. From a practical viewpoint, the effectiveness of any material is related to only the total volume of micropores (less than 2 nm diameter) it contains, as distinct from the larger meso and macropores. Pores greater than 2 nm are ineffective because adsorption is not accompanied by capillary condensation. The aim therefore is at producing a material containing a high proportion of micropores, and in a form having as large a surface area as possible. Adsorption increases with pressure until, at a certain limiting value, the rate of increase flattens off. For this reason, the pressure chosen in practice for storage of natural gas in activated carbon is about 35 bar.

Of particular interest is that the density of the molecules adsorbed into micropores is similar to that of the gas in its liquid state, in which 600 volumes of gas under standard atmospheric conditions become one volume. Of more practical relevance is the fact that, at 35 bar, the density of the gas stored by adsorption in materials currently commercially available is equivalent to that of the gas compressed at 100 bar in a conventional pressure vessel. Future developments in adsorbent manufacture could increase the figure for the equivalent pressure to between 150 and 170 bar.

During the charging of the carbon pack, its temperature rises, the highest temperature occurring in the centre owing to heat transfer from the outer layers to the surroundings. During discharge, a fall in temperature is of course experienced. The temperature range in the centre of the pack depends on factors such as its size, changes in gas pressures, the rates of charge and discharge, and the rate of conduction to the surroundings. At an ambient temperature of 25 deg C, the range could perhaps be from 70 deg C to -40 deg C. Consequently, the rate of filling (either overnight or rapid refuelling) will influence the capacity of the carbon pack. At the Institute of Gas Technology, Chicago, materials undergoing phase change at or near ambient temperature have been introduced into the carbon pack, thus obtaining what is claimed to have been near isothermal operation. It is undesirable for the temperature to fall below about -5 deg C in the vehicle, otherwise the gas supply might be inadequate for sudden acceleration.

Work done so far has indicated that, if hydrocarbons higher than C_4 are present in the natural gas, a degree of irreversible storage occurs, reducing the capacity by 20–30% over 100 cycles. Other work, but with Canadian natural gas, has indicated a levelling off after 40 cycles, with no further deterioration in storage capacity. With pure methane, on the other hand, no deterioration has been experienced over 100 cycles.

So-called activated carbons comprising volumetrically high proportions of micropores have been produced for hundreds of years. Among the most suitable are those made from coconut shells. In the form of granules and powders, however, they are not ideal because too much of their volumes comprise intergranular voids. Work by the London Research Station, working with Sutcliffe Speakman Carbons, of Leigh, Lancashire, has demonstrated that the best results are obtained by compacting the activated carbon powder to produce mouldings, for example in the form of discs about 30 mm thick or more. In this form, the surface area exposed to the gas is of the order of 1500 m^2 per gm and the density of the compacted carbon 0.8–1.0 gm per cm^2. Indeed, stacked in cylindrical containers, the discs can offer a much higher proportion of micropores per unit volume than either powder or granule fillings.

3.9 Safety of CNG

When, in July 1990, a devastating fire destroyed all 35 buses in a Central Netherlands Transport Garage, Utrecht, two experimental natural gas fuelled vehicles happened to be there. Though destroyed with the rest, the natural gas fuelled vehicles, despite storage pressures rising from about 240 bar, contributed less to the overall conflagration than any of the diesel fuelled vehicles.

This was because, in addition to the electromagnetically controlled shut-off valve used also in LPG fuelled vehicles, the LNG tanks had a burst-disc fitted in association with a restrictor valve to limit the rate of outflow of gas through the aperture from which the disc was blown. Consequently, when the heat of the fire increased the pressure, blowing the disc out, the gas was discharged at a controlled rate. The jet of gas thus released burned steadily but without exacerbating the fire hazard, and the cylindrical tanks remained intact. This is in stark contrast to what has happened, for example, in multiple pile-ups on motorways, where bursting fuel tanks have sprayed burning liquid fuel over other vehicles and set light to them. An alternative to the burst-disc is a wax plug which will melt in the event of a fire. British Gas, in its experimental car installation, fitted both a burst-disc and a wax plug.

In very hot climates, there is a risk of increases in pressure due to rises in ambient temperature. The problem generally arises if the tank is refuelled at night when it is cold, and the temperature becomes very high during the following day. To cater for such conditions, a special valve can be fitted which, after allowing the gas to blow off, closes again as the pressure inside falls to its normal maximum.

Petroleum gas is heavier than air and, if it leaks, tends to fall to the floor and to run on down into maintenance pits, underfloor service channels and cellars. Consequently, a lighted cigarette or even a spark from a nail in the heel of a boot can ignite it and cause an explosion. For this reason, legislation relative to bulk storage requires the tank to be in the centre of a large area of land, to reduce the danger in the event of a fire.

Natural gas, on the other hand, is much lighter than air (about half its density) and therefore tends to dissipate upwards either into the open air or through roof vents in a garage, and thus rapidly moves out of harm's way. Another consideration is that, while LPG will form an ignitable mixture with anything between 1.5 and 10 times its own volume of air, natural gas has the very narrow flammability range of only 4% to 15% of its own volume with air. Incidentally, in terms of mass, the flammability of natural gas is 2.3–8.9%. Moreover, its flash point, at 700°C, is approximately twice as high as that of gasoline, and the energy required to ignite a combustible mixture is much greater too, hence its low fire risk.

3.10 Other advantages of natural gas

The little sulphur that natural gas contains can be easily removed, so it can be delivered in a largely unpolluted state to the user. Evaporation is not a prerequisite for mixture formation, so a vaporiser is not needed between the tank and the the engine induction manifold, as it is with LPG, nor is enrichment necessary for running under cold conditions. Consequently, this

Table 3.5 EMISSIONS FROM THE NG L10 ENGINE AND THE 1991 LEGAL LIMITS

Emission	NG L10, g/bhp h	1991 standards, g/bhp h
NO_x	4.5	5.0
CO	<5.0	15.5
HC	<1.3*	1.3
Particulates	0.05	0.1

* With an oxidation catalyst

gas burns very cleanly, even at very low engine temperatures, and good economy and very low HC emissions are obtained by comparison with those from either a gasoline or diesel engines. Natural gas has an Octane rating between 120 and 130, depending on its precise composition, so high compression ratios and therefore high efficiencies, or miles per gallon, are attainable. In fact, according to Steve Gauthier, of the Gas Research Institute, compression ratios of 15:1–18:1 are usable Moreover, because the gas burns so cleanly, the life expectancy of not only components such as plugs, filters and piston rings but also the engines themselves can be expected to be doubled.

As can be seen from Table 3.5, the exhaust emissions obtained with the US Gas Research Institute's NG L10, 10 litre 240 bhp dedicated natural gas engine have been shown to be within the limits laid down in the EPA 1991 standards. The maximum thermal efficiency of this engine was 37%. Methane is non-reactive and so does not itself encourage photochemical reactions. HC emissions are low because of the absence of both evaporation and the need to enrich the mixture for cold starting and low temperature running. In any case, given a stoichiometric mixture, the gas burns more completely than evaporated liquid hydrocarbons. Moreover, in the USA, the EPA estimates that, by using natural gas, as compared with liquid hydrocarbon fuels and, with the application of lean-burn technology, well over a 50% reduction in CO emissions and even greater reductions in NO_x. could be obtained. Application of lean burn technology is expected to reduce NO_x levels below those of conventional combustion.

3.11 Operational experience with natural gas

Of the 70,000 natural fuelled vehicles in regular service, most are in Italy, New Zealand and Russia and the USA. In the UK, British gas have been doing development work with a Volvo Estate, and two Ford Escort vans. Also, a Transit van and a Maestro have been converted in Holland for further work in England. A breakthrough, might be made initially with commercial vehicles, including buses, used on scheduled local runs, though taxi operation is another prospect for this fuel.

Because of the relatively low flame speed of natural gas in air, the ignition has to be advanced relative to the settings used for gasoline. Possibly a high energy spark would be an added advantage. High energy systems are in any case used currently for gasoline engines with closed loop control, to avoid any possibility of overheating of the oxidation catalyst, so this does not present any difficulty. With engines designed for unleaded fuel, it seems unlikely that there will be any significant valve sinkage problems.

In its development work, British gas are using a 15.25 cm (6 in) diameter cylindrical tank, made of aluminium to keep the weight as low as possible. However, they have also considered tanks of annular, or doughnut, form and, on the same principle, flat rectangular tanks with vertical holes through them, the walls around the holes acting as cylindrical ties to prevent the tanks from ballooning under pressure. The latter are 1 m by 1 m by 0.165 m thick and hold 106 dm^3 of gas.

Incentives for the use of natural gas include its current price, which is the equivalent of about £1.50 per gallon. Conversion kits currently cost about £750 but, in the future, costs could come down with large scale production and even more so if vehicles were to come off the production line equipped for operation on gas.

As regards bulk supply, it could be piped directly to individual distribution points, though only at low presure. At each point, all that would be necessary would be a simple installation comprising a compressor, such as are supplied by Sulzer Burckhardt, to serve the metering equipment, from which could be taken one or more flexible delivery hoses with dispensing guns on their ends. The delivery nozzles for either LPG or CNG contain automatic check-valves and have screw-on connectors on their ends.

An alternative might be daily deliveries of a load of charged tanks in exchange for empties. Obviously, some ingenuity would have to be applied in the design of durable connectors. In the context of safety, the regulations applying to the bulk storage and distribution of LPG could hardly be satisfactory for natural gas, so we should be looking at the experience embodied in the New Zealand and American regulations as a basis for new ones for Europe.

New reserves of crude oil, natural gas and coal are constantly being found, so it is difficult to quote ultimate figures. However, reserves of natural gas have been variously estimated at more than double those of crude oil. Consequently, natural gas could have a major part to play in delaying the point at which all our fossil fuels will all be exhausted. It is already being used as a fuel for vehicles in countries where it is to be found in large quantities underground.

Chapter 4

Transferring the fuel from tank to engine

Fuel tank layout, construction, materials, filling and venting have been dealt with in great detail in Vol. 1, so little more remains to be added. In diesel fuel systems, it may be desirable to form in the base of the tank a sump into which any water will fall and be kept well clear of the lower end of the fuel pick-up so that, if it freezes, it will not block the fuel supply. On the other hand, some authorities hold that, in such a sump, biological growth can occur. Consequently, they believe it is better to draw the fuel through a stand pipe to avoid the problem of freezing, and fit a drain cock to the tank, so that the water can be cleared periodically.

This could be the better solution for farm tractors, which may stand unused for long periods, and the risk of biological growth is in any case, more significant. For road vehicle applications especially, however, there are two objections to this arrangement. First, operators generally tend to regard their tanks only as fuel containers, and therefore tend not to take seriously instructions regarding drawing water off. Secondly, it is usually less costly to insert water separator units into the fuel system than to produce fuel tanks designed to serve that purpose.

Pipes from tank to lift pump, too, were dealt with in Vol. 1, but with diesel engines, the potential for wax to form in the fuel at low temperatures is an additional major consideration. Bends of small radius should be avoided and the pipes should slope continuously upwards from the tank so that air cannot be trapped in them nor water settle in loops, where it could freeze and block the system.

In many of the fuels commercially available, the potential for wax to form at low temperatures is such that it could do so in almost any size of pipe that might be reasonably considered. Some vehicle manufacturers rationalise on a single bore size of about 6 mm, for low pressure pipes. In this respect, much depends upon the type of installation and, of course the power output of the engine. On the other hand, Lucas design their systems on the basis of a viscosity of 15 cSt for flow through pipes.

4.1 High pressure fuel pipes

The requirements for the high pressure pipes delivering the fuel from the injection pump are much more complex. The runs should be as short, straight,

Fig. 4.1 This one-piece plastics pipe-clip for four tubes has an integral hinge at one end and inter-locking barbs at the other to hold it closed when snapped shut. As it closes, the wedge shape blades enter the gaps between the individual clasps to press them tightly around the tubes

and stiff as practicable, to avoid excessive vibration, and the bores smooth. Bends, where essential, should be of as large a radius as possible, and the pipes should be supported in a manner such as to form nodes at short intervals along them, so that only high frequency low amplitude vibrations can occur, preferably outside the range of the forcing frequencies at normal engine speeds. In general, the straight lengths between each end fitting and the nearest bend should be short, to ensure that the natural frequency of vibration of this section is well above the normal frequency range of the engine vibrations. Clamps of rubber lined metal strip are being supplanted by plastics fittings that are slid into T-grooves in the components to which the pipes are to be secured and then snapped shut around a tube or bunch of tubes, Fig. 4.1. The number and positioning of clips, to create nodes and thus lift the natural frequencies outside those of the engine, is of course critical for limiting amplitudes of vibration.

Because these pipes are subjected to violently fluctuating pressures which, in modern engines, can be of well over 1000 bar and with superimposed shock waves, major considerations are strength and cavitation. As previously indicated, by using pipes of adequate stiffness and supporting them at intervals along their lengths, resonant frequencies of radial and bending vibration respectively can be raised above the forcing frequency range of the engine. Cavitation damage is avoided by ensuring that the bores of the pipes are smooth and are protected against corrosion, the radii of bends are as large as practicable, and there are no discontinuities of bore between the pipes, their end fittings and the parts of the pump and injectors to which they are connected. At the bend, ovallity in excess of -1.0 and $+0.2$ mm should be avoided.

Pipes of a wide range of materials have been used. Most commonly, however, mild steel, zinc or tin plated for protection against corrosion, is employed The steel is first hot rolled and then cold drawn to produce a seamless tube. During the cold drawing process, a very hard alloy steel ball is pushed through the tube to size its bore accurately. High quality material is essential for the avoidance of high levels of scrap, since even the smallest of inclusions can be greatly elongated by the extrusion process and lead to fatigue failures or even seepage of fuel through the wall. Scrap therefore can be excessive. The harder the material, the greater is its resistance to cavitation erosion so, following the introduction of very high injection pressures for reducing undesirable exhaust emissions, stainless steel pipes have been actively under consideration.

A high standard of cleanliness in manufacture is essential if warranty costs arising from damage by debris are to be avoided. Moreover, pipes must be produced with bore diameters to close tolerances, otherwise the cylinders might receive unequal quantities of fuel, which could cause uneven running and adversely affect fuel economy.

The majority of engine manufacturers have 6 mm outside diameter tubing, and vary the inside diameter, from about 1 to 3 mm, to suit the application. If injection pressures are low, the thinner wall is both strong and stiff enough; if they are high, the smaller bore is still adequate for delivering the quantity of fuel needed. By virtue of a standard outside diameter, pipe fixing clips, union nuts and other fittings can be rationalised.

In addition to their plain cold-drawn mild steel tube, Bundy, a Division of the TI Group plc, produces multi-layer tubing for high pressure diesel applications. As can be seen from Fig. 4.2, it comprises three parts: two are concentric tubes, termed *double wall brakes*, of copper coated steel and the third a seamless tubular liner. All three are assembled together and then reduced by *sinking down* (drawing down) to a high precision finished size.

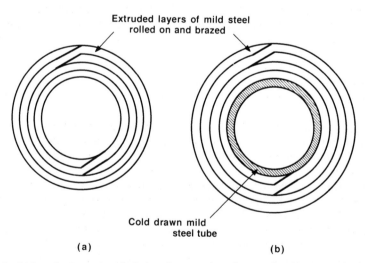

Fig. 4.2 Right, tube the section of which is shown on the right is produced by TI Bundy, for high pressure diesel injection systems. The section on the left, without the cold drawn seamless liner, is not recommended because it may suffer erosion of the feather edge in the bore, by the fuel flow and pressure waves

The whole assembly is then passed through a furnace in which all five layers are brazed together.

This type of tubing has the advantage of great strength and rigidity together with a useful degree of internal damping, for reducing amplitudes of vibration. Moreover, by virtue of of its multi-layer construction, it offers greater security under conditions of severe fatigue loading, when imperfections in cold drawn materials could lead to leakage.

To obtain a higher quality finish in the bore of the finished product, the liner is more finely drawn down than conventional cold drawn tubing. The higher the pressures the greater are of course the risks of cavitation damage, so Bundy match the hardness of the liner to the application. This, they claim, is more cost-effective than resorting to stainless steel tubes to cater for high pressures.

4.2 Pipe end fittings and connections

Detail attention to the design of the end fittings and connections is essential for the avoidance of leakage and fatigue failures. The types of component used will depend on the size of engine, fuel pressures, modes and amplitudes of vibration likely to be experienced, and rate of fuel flow. Some manufacturers bell-mouth the end of the tube and seat an olive in it. This is suitable for the lower injection pressures, and has the advantages of simplicity and self-alignment potential but, if the pipe and component to which it is to be connected are badly out of line, damage can occur. What happens is that one side of the end of a badly aligned pipe touches and then is pulled hard down on to the component to which it is connected. This can cause high local stresses, raise burrs and even start cracks, all of which can lead to stress concentrations and therefore fatigue failures.

In service, mechanics handling the olives with dirty fingers could leave particles of foreign matter trapped between the sealing surfaces. Moreover, olives are easily dropped and lost during dismantling and assembly operations.

An alternative of a hemispherically ended fitting brazed on to the pipe can be handled without touching the sealing surface and therefore tends to be less liable to dirt entrapment. The most common arrangement, however, is to bump up the end of the tube to form the spherical seating.

End fittings can be formed or simply secured by bumping up the end of the tube, Fig. 4.3(a). Essentially, the process involves inserting a short mandrel into the bore, to ensure that it is not closed at all by the bumping up operation. A chuck around the end of the tube holds it securely while a tool applies a pressure to its end, either to form it into the required shape or to lock it into a sleeve component, as in Fig. 4.3(b). The mandrel should be slightly tapered so that it will not scuff the bore as it is being withdrawn, but not so much as to cause a discontinuity of fuel flow through the bore.

Prior to forming and bumping up the ends, cold drawn tube has to be

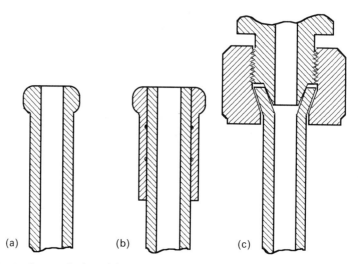

Fig. 4.3 A selection of tube end-fittings

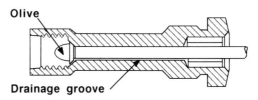

Fig. 4.4 This tube end-fitting is used by the Ford Motor Company, for applications in which the injectors are totally enclosed by the rocker cover. The left-hand end is inserted through an aperture in the cover and screwed on to the injector. The groove beneath the head at the other end carries a sealing ring

annealed. Given suitable equipment, it may be practicable to anneal only the parts of the tube that will be formed but, in either case, the tin coating has to be done afterwards. In some of the examples illustrated, sealing is effected by brazing the sleeve to the tube.

To obviate the problem of local crushing of the edge of a misaligned tube on one side of the connection, tapered, instead of spherical pipe flares and seatings have been used. If injection pressures are very high, however, a smaller area of seating, and therefore higher seating pressures and more secure sealing are obtained by making the angle of the female taper slightly more acute than that of the male part, Fig. 4.3(c). Where the injector is enclosed within the camshaft cover, it is important to ensure that, if any leakage occurs, it is discharged outside the cover, otherwise serious dilution of the lubricating oil could arise. A method the Ford Motor Co. uses to avoid this is illustrated in Fig. 4.4. Should any fuel leak past the seating, it passes along a groove machined in bore of the sleeve and out at the other end, which is outside the camshaft cover.

Some of the joints already described can be made more reliably fuel-tight by the application of a sealing compound. A point to bear in mind is that the liquid sealant must not be allowed to leak away into the pipeline and pass on to the injectors, so only a slight smear should be applied. Loctite anaerobic sealant is particularly suitable because it sets hard as soon as air is excluded from it. When the spherical or tapered parts are pulled together, the liquid is squeezed from between the faces until metal-to-metal contact is made, the only sealant remaining being that left in surface depressions where metal fails to contact metal. In other words, it ensures that what might otherwise be an incomplete seal is in fact complete. An advantage, as compared with elastomeric seals, is that once the joint has been made, no further settling can occur, so there is no possibility of relaxation of the joint.

Concentricity of end fitting and component to which the pipe is attached is another important requirement. The best practice is to machine both the threads to which the union nut is screwed and the bore in the component simultaneously. To ensure that no debris is left adhering to the bores of the pipes, ultrasonic cleaning is recommended. Caution has to be exercised with wrapped tubing that does not have a cold drawn seamless liner. With such components it is necessary to ensure that the ultrasonic vibration does not feather the edges of the inner wrapped strip.

4.3 Fuel conditioning

For diesel engines, good filtration is critical. Four types of contamination have to be guarded against: organic sludge, inorganic abrasive debris, water and wax crystals. Debris due to either corrosion or wear can cause havoc in both pumps and injectors.

The clearances between the pump plungers and barrels and between the moving parts of the injectors are so fine (generally 1 to 2 μm) that particles of microscopic size will cause wear, scoring, or even seizure and severe damage to the mechanism. Furthermore, wear can occur as a result of the shearing of the abrasive particles between the closing edges of the plungers and the delivery and spill ports (Chapter 5). Failure of injectors to seat properly, whether due to debris on the seats or sticking of the moving parts, can cause dribbling and, consequently, carbon build-up leading even to total blockage.

Distributor type pumps are more sensitive than the in-line type to the presence of abrasive particulate matter, debris in general and corrosive influences in the fuel. They generally function satisfactorily with only single stage filtration in cars. For off highway vehicles, on the other hand, because they operate in conditions in which gross contamination of fuel is liable to occur, a sedimentor is generally interposed between the tank and feed pump. This, together with a fine filter capable of arresting particles down to 5 μm diameter between the feed pump and engine, is adequate for removing the harmful substances from the fuel.

Under very cold conditions, the choking of filters by ice or wax can be avoided by incorporating electric heater elements in the water separators and filter assemblies. In both filters and combined filter and water separator units, the heating element must always be interposed before the filter element. Since, in sedimentors, the direction of inflow is generally downwards, the heating element needs to be at the top. In a few applications engine coolant water heating elements are employed.

Some typical fuel circuits for distributor type pumps are illustrated in Fig. 4.5, and for an in-line pump in Fig. 4.6. The two filters are separate units mounted on brackets on either the vehicle structure or the engine itself. However, the screw-on cartridge type is available as an alternative and can be screwed into either a special head or directly to the injection pump.

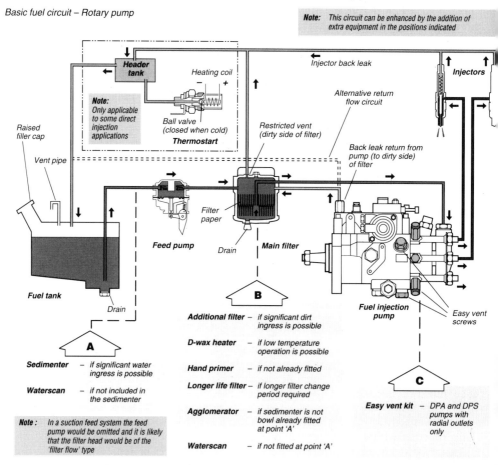

Fig. 4.5 Comprehensive recommendations by Lucas for the design of the hydraulic system for a rotary distributor type injection pump

TRANSFERRING THE FUEL FROM TANK TO ENGINE 43

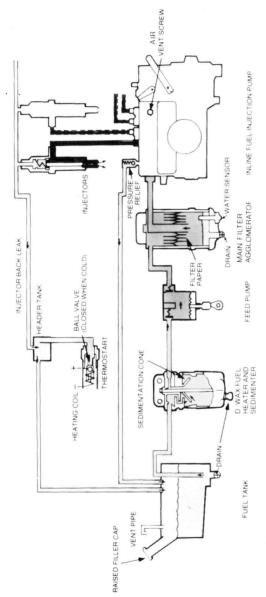

Fig. 4.6 Typical fuel circuit recommended by Lucas for an in-line type injection pump

4.4 Solid particles and sludge

Dust can be prevented from entering the fuel in service by fitting an an elementary filter over the upper end of the tank vent. This is essential for vehicles that operate in dusty conditions, especially farm tractors for example. Similarly, the size of the the vent should be severely restricted because, with the fuel swilling and slopping around inside the tank, there is a tendency for air, together with any water and dust it may contain, to flow continuously in and out.

For protection of the injection equipment from dust, a filter or filters must be incorporated in the fuel feed system. A fuel filter capable of blocking the passage of all particles, regardless of size, would be too large and too costly to be acceptable. Lucas Diesel Systems consider that, on a car, a filter capable of passing at least 1000 Imp. gal (4546 litres) of fuel before choking would last for about 50,000 miles, though the actual figure does depend upon the fuel consumption.

During its useful life, a filter should retain at least 80% of particles of about 7 μm diameter (ISO 4020). The useful life is a function of the area and porosity of the filtering element. Obviously the greater the area of this element the longer will be its useful life. Therefore, paper elements offer the best compromise between particle retention and long life. The type of paper used, its porosity, the method of production and treatment of the fibres, such as resin impregnation to bond them together, are of course critical. In many instances, the paper is creped to ensure that there is adequate fuel flow path between the layers of paper.

Elements providing filtration areas of 3 cm^2/cm^3 volume are normal, and 10 cm^2/cm^3 are practicable. They generally take one of three forms: pleated, or star shape elements; coiled V-construction; or very rarely, spaced stacked discs. The first two are so commonplace in all types of filtration as to need little further explanation. In the case of star section elements, the core tube has to be perforated. Felt filters are now rarely to be found in diesel engine fuel systems.

Coiled V-section elements are made by either folding strips of the paper or, more usually by bonding alternate edges and, in each case, coiling them into cylindrical form around a central tubular core. If they are folded, the edges have to be crimped, otherwise they will not roll compactly and uniformly on to the core.

Setting standard tests for filters has been difficult because current fuel specifications do not include any reference to filter choking effect. Worldwide, the choking effect varies by a factor in excess of 10:1. Life until choked is more dependent upon the presence of organic materials in the fuel than on the quantity of particles of inorganic material trapped. If a filter is tested with a fuel that contains virtually no organic choking materials, the particles build up on it, forming a porous bed, which progressively improves its retention capability but without necessarily choking it, but this is not a condition that occurs in service.

Determination of the time between filter changes is done by taking a fuel

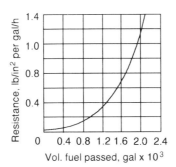

Fig. 4.7 Results of a filter test, using fuel of lower than average cleanliness and quality, to determine the life to be recommended between element changes

of lower than average quality, as regards organic choking material content, that is available in the area in which it is to be used, and testing it to determine the rate of choking as indicated by filter resistance plotted against volume of fuel passed, as shown in Fig. 4.7. Filter element change intervals are usually those judged by the engine and injection equipment manufacturers to be reasonable, but without significantly penalising users of good quality fuel. If filters choke before a change is due, the cause is usually heavier than normal contamination of the fuel in storage.

Under certain particularly severe operating conditions, short replacement intervals may be found necessary, in which case consideration should be given to installing either a larger filter or two in parallel, to obtain a larger effective area for filtration. If two filters of the same size are fitted in series, the element of that which is upstream will of course need to be changed more frequently than that in the downstream filter. Even so, in rare circumstances or exceptional conditions, such an arrangement may be desirable.

4.5 Removal of water

Even given the degree of care normally taken in both keeping water out of storage facilities and draining off that which does collect there, some inevitably gets into the fuel tank on the vehicle owing to both condensation from the air in it and precipitation from that already in solution before delivery from bulk storage. A curve of water solubility in diesel fuel is illustrated in Fig. 4.8. Obviously, fuel tank vents have to be shielded to prevent both rain and spray from entering. Even so, some rain will drop in to the tank while it is being filled in wet weather. According to BS2869, the quantity of water in the fuel for high speed diesel engines should not exceed 0.05%. Under poor conditions of storage, it can exceptionally creep up to as high as 2% At such a high level, though, the remedy is positively the responsibility of the fuel supplier rather than the vehicle or its fuel system designers and suppliers.

Precipitation occurs daily towards nightfall, as the temperature drops.

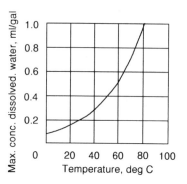

Fig. 4.8 Solubility of water in a typical diesel fuel

Flexible skins, probably due to the deposition of asphaltine micelles, form around the precipitated drops of water, and their thickness tends to increase with time. As the droplets settle down in the base of the tank, the skin re-forms on the surface and tends to prevent the water from redissolving as the temperature rises again. An important consideration is that the water may have dwelt, in both the fuel tank on the vehicle and bulk storage, long enough for sulphate reducing bacteria to have multiplied at the fuel-water interface and generated acid. Another possibility is that the water may be contaminated with common salt, which of course will accelerate corrosion.

With or without acid, however, the water could damage the pumping elements in two ways. First, owing to the inferior lubrication properties of water as compared with diesel oil, its presence can lead to scuffing and rapid wear. Secondly, trapped between the rubbing surfaces of the elements overnight it can cause corrosion. Contamination with water can also cause pitting of the cams that actuate the pumping plungers, failure of springs and rusting of other steel components.

Combined filter and water separators are available. In these, the fine droplets of water are agglomerated in the filter element to form larger drops which fall into the base of the unit. Agglomeration in such units always occurs on the clean side of the filter.

Among the simpler primary water separators, for fitting upstream of the pump, is a barrier type vertical flow sedimentor. In this the fuel, after being delivered into the sedimentation bowl, passes vertically up to disc of very fine gauze or some other thin filtering medium. This traps any water droplets in it, which then drop down to the base of the bowl.

In other sedimentors the velocity of flow is reduced simply by passing it through a plenum chamber, or bowl, to allow the water droplets similarly to fall to the bottom. They can take either of three forms. One is the horizontal, or linear flow, type in which the contaminated fuel flows across the separator bowl. In this type, the fuel flow is straightened by diffuser vanes as it leaves the inlet and, again, as it approaches the outlet, so that it is spread as uniformly as practicable over a horizontal cross section of the bowl.

Fig. 4.9 A Lucas sedimenter with a Waterscan electronic water level detector screwed into its base. The drain plug is on the opposite side of the base

In the second category are the radial flow types, in one of which the fuel enters at the centre and flows radially outwards over a disc or conical baffle, Fig. 4.9, before dropping down through the clearance around its periphery into the sedimentation bowl, where the water drops to the bottom. From here, the cleansed fuel passes radially inwards again and then up to the outlet port. This arrangement has the advantage of simplicity and compactness, since both the inlet and outlet ports can be incorporated in the filter head casting.

In another form of radial flow type, the contaminated fuel enters through a toroidal gallery approximately mid-way between the top and base of the bowl. As it flows radially towards the centre, before passing down through the upper end of a stack pipe, the water agglomerates and sinks to the bottom. From the stack pipe the cleansed fuel passes on to the feed pump. This type is more suitable fore industrial than automotive applications.

The effectiveness of all types of sedimentor is a function of the horizontal cross sectional area of the bowl, and all can be designed for removal of water droplets down to 100 μm. They are fitted on the suction side of the feed pump. This is because, after the fuel has passed through the pump the delivery side, the water will have have been broken down into droplets so tiny that they will not so readily fall to the base of sedimentation bowl.

Clearly, in these sedimentors, droplets below a certain diameter will fail to drop down through the fuel into the base of the bowl. This limiting diameter is given by a formula as follows, based on Stoke's law:

$$d = 1.35 \times 10^3 \frac{\sqrt{q\eta}}{\sqrt{A(\rho_w - \rho_f)}}$$

where

A = effective horizontal area of the sedimentor, cm^2
d = cut-off size of the water droplets, μm
q = rate of flow of the fuel, ml/sec
η = viscosity of the fuel, poise
$(\rho_w - \rho_f)$ = the difference between the densities of water and fuel, g/ml.

Solid particles, too, will sediment at a rate that can be calculated the basis of Stoke's law, in the bowl. Those remaining in the fuel will be trapped in the filter downstream. The minimum size of particles thus trapped will of course depend on its material and construction.

Provision is made for draining the water off periodically. With an electrical high water-level sensor in the base of the sedimentation bowl, a warning lamp on the dash fascia can be illuminated to indicate to the driver that he should drain off the water in the bowl. Alternatives are a float type water level indicator or, by fitting a transparent bowl, to provide for visual checking. It would possible perform the drainage function automatically, though such a system would be costly. However, owing to the risk of depositing fuel on the road and thus causing its surface to become lethally slippery, such a system would be unsuitable for automotive applications. To guard against the discharge of fuel into the atmosphere a low water-level sensor would have to be incorporated in the control system.

Centrifugal separators are ruled out owing to their high cost and the need to provide them with a drive. In general, it is advisable for sedimentors to be installed as close as practicable to the fuel tank, to obviate all possibility of water freezing in the fuel pipes and blocking them.

4.6 Removal of wax

Small crystals of wax formed in very cold weather, as explained in Chapter 2, can adversely affect cold operation. Under extreme conditions, the engine may not even start at. However, more commonly, it starts and runs for perhaps 5 or ten minutes, while the wax crystals build up until they block the pipeline or clog the filter, and then stalls and cannot be started again until the fuel temperature rises above the wax precipitation point. The fuel filter is generally the critical area as regards blockage by wax.

Where extremely low temperatures may be experienced, a better course is to incorporate a fuel heater element, either in or upstream of the filter. This enables the driver to warm the fuel before he starts the engine. Incidentally, because of the risk of blockage by wax crystals, it is inadvisable to have a wire mesh strainer on the lower end of the fuel pick-up in the tank.

Given fuel of reasonably good quality, wax precipitation can be avoided simply by mounting the filter close to the engine and fitting radiator blinds. The cold filter plugging point specified for disel fuel depends upon local climatic conditions. For example, in Finland it is about -30 deg C, while in Singapore it may be as high as 8 deg C.

4.7 Filtration arrangements

A typical installation of primary and secondary filters comprises a primary filter combined with or, much less commonly in addition to, the sedimenter. The main function of the primary filter is to protect the feed pump from the effect of harmful abrasives. Because it is on the suction side of the pump, it is required to offer a minimum resistance to flow, and therefore is fairly coarse, otherwise fuel starvation could occur.

The secondary filter of course removes the very fine particles that have passed through the primary. Its paper element therefore has a very fine and dense structure. For stationary engine applications, low readings on a fuel pressure gauge on the outlet from this filter can be used as an indication that the element needs to be changed.

A manual primer pump is often fitted to the filter head, as in Fig. 4.10. It is normally used after element changes or any other maintenance operation that lets air into the system. When the engine is running, the fuel should preferably bypass the primer pump, to avoid any possibility of restriction to flow. Venting procedures for removal of air vary from installation to installation. However, some pumps are self-venting, in which case most return the air with the excess fuel to the tank. There is normally a non-return valve in the vent.

Electric heater elements the ratings of which are dependent on size of engine and type of application, but usually of about 200–300 W, may also be incorporated in the filter to prevent it from being clogged by wax. Thermostatic control is desirable, to prevent overheating of the fuel. An alternative is to use a negative coefficient heating element, in other words, one having a resistance characteristic rising either progressively or sharply at a certain critical temperature, thus limiting the heat output of the element. The primary filter is the usual site for a heater but, in some vehicles, it may be further upstream.

If it is in the combined filter and water separator, the heater coil may either surround the inlet or be installed immediately above or beneath the filter element. It can be argued that, in the latter position, the heat rises to warm the filter element, while it is also well sited for melting any ice that may have formed in the water separator below. However, most manufacturers hold the view that the heating should be applied above the filter element.

4.8 Some production filters and water separators

Fine filters for secondary application are generally of simple design, comprising a cast head in which the inlet and outlet passages are cored.

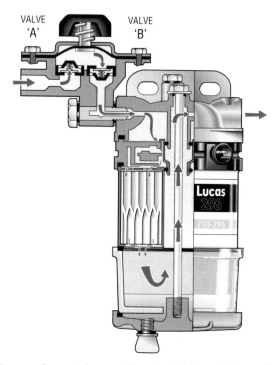

Fig. 4.10 This Lucas agglomerater houses a filter element, below which is a sedimenter base, and a manual priming pump is screwed into the filter head. The inlet valve is at A and the outlet at B. This unit can be installed in series with a sedimenter, which may be connected directly to its inlet

Mounted beneath it and usually secured by an axial bolt or nut is the filter chamber. This houses a cylindrical filter element, of either the pleated or star section, stacked spaced disc, or coiled V-section paper. If all water has been removed by the primary filter, it may not be absolutely essential to have a water trap and drain plug in the base of the secondary filter, though many manufacturers prefer to play safe and incorporate one.

In the pleated star and stacked disc types of filter, the fuel is generally directed down outside of the filter element and, after passing radially inwards through it, rises up a central tube to the outlet. The more ingenious V-section paper type, is different. Fuel passes vertically down into the folded V-sections and then radially through the filter paper, as shown in Fig. 4.11.

Technocar, of Turin, produce the unit illustrated in Fig. 4.12. Incorporated in its head are a manual priming pump and a heater assembly. The latter comprises five extremely compact ceramic based elements the electrical resistance of which increases logarithmically with temperature. On starting the engine, the current through the elements is 35 A, but decreases logarithmically within the first 60 sec to about 10–12 A. An electronic control switches the heater off altogether if the temperature of the fuel is above 2 deg C.

An agglomerator type diesel fuel filter unit produced by Lucas is illustrated

TRANSFERRING THE FUEL FROM TANK TO ENGINE

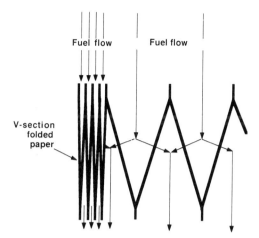

Fig. 4.11 Diagram illustrating the flow through a folded, or bonded, V-section paper filter element

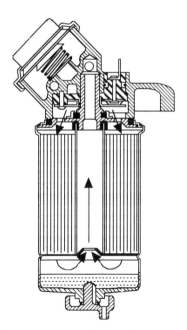

Fig. 4.12 An automatically varying resistance heating element is installed in the Technocar filter head, in which is also incorporated a priming pump

in Fig. 4.10. Its filtration element comprises two strips of crepe paper wound on to a tubular core, which is contained co-axially within a cartridge or metal canister. One of the strips is coated with the bonding agent along one edge of one side and the other edge of the other side prior to the winding operation, during which they bond together to form a multi-V section, as illustrated in Fig. 4.11. Crepe paper is used to provide a flow path between the layers of paper.

A cylindrical glass water-separator fitted beneath this filter element can be drained by removing a winged screw plug from a hole in its metal base plate. The whole assembly is held together by a single bolt inserted downwards through a hole in the centre of the head, into the upper end of a retainer stem which, in turn, is screwed into the base plate at the bottom of the water separator. Sealing rings are fitted in the joints between the head and filter housing, and between the water separator and base plate. As can be seen from the illustration, a one-piece casting can be easily substituted for the cylindrical glass separator and its base plate. Bosch supply a similar filter, but it comprises a single strip of paper folded to form the V-section, and the bonding agent is applied along the outer faces of its edges at the open end of the V. To ensure that it will wind snugly and smoothly on to the core, the folded edge is crimped.

Fuel enters through a radial port into a chamber in the head of the Lucas filter. It then passes down through perforations in the top plate of the filter element and out through holes around the periphery of the lower plate. On the way down, therefore, it has to pass both axially between and radially though the folds in the crepe paper. Any very fine droplets of water remaining in the fuel, after it has passed through the primary filter, agglomerate and settle out in the separator bowl below. Sludge and solid particles are trapped and contained in the creped filter paper. The cleansed fuel leaving through perforations in the lower plate of the filter element passes down into the water separator, and then up through the clearance between the core tube and the retainer stem, and out through another port radially opposite the inlet.

Either a water separator, generally similar to the Lucas Waterscan unit, illustrated in Fig. 4.9, or a primary filter/agglomerator is interposed between this filter and the tank. Waterscan is simply a water trap available with either a transparent or an aluminium separator bowl, with a drain screw in its base. Fuel entering it passes over a sedimentor cone and down through the gap between its periphery and the walls of the bowl. Water and sediment fall down to the base of the bowl and the clean fuel passes up and away through an axial passage in its centre. It has an electronic probe in its base, to warn the driver that level of water is too high. This can be of either the capacitive or conductive type. The conductive probe has two electrodes. As the level of water rises, the balance of the system is disturbed, giving a signal which triggers a lamp, buzzer or other suitable warning device in the driver's cab. Its three-wire electrical circuit has connections to earth, the battery positive terminal and the warning lamp.

To cater for very cold conditions, Lucas offer their D-Wax 150 or 300 W unit, which is an electrical heater available as either original equipment or for retrofit. This can be interposed between the head and body of a wide range of their filter and water trap products.

4.9 Fuel heating for operation in very cold climates

In countries where severe cold is the norm, rather than occurring a few days each year, something better than an electric heater in the filter might be desirable. Mostly, heaters or heat exchangers around the fuel pipes or in the tank are too costly and bulky to be practicable. However, if a diesel or paraffin fired heater and heat-exchange system is in any case installed to warm up the engine coolant system and sump, or both, the cost of taking water pipes to the fuel tank is not so great, especially if the heat exchange element is simply a coiled part of the water pipe.

Eberspächer supply just such a system, which can also be used for heating the driver's cab, Fig. 4.13. The heat-exchanger coil assembly for insertion into the tank is called the Arctic Fox. A 5 kW heater warms water piped from the engine coolant system. It has its own circulation pump, and therefore can be set in operation long before the engine is started. A 24 hour digital timer can be included in the equipment, so that it can be fired automatically at a preset time.

The advantage of such a system is that it is energy efficient. Moreover, the whole tank full of fuel is brought up to the required operating temperature and, since the quantity of heat it contains is such that it takes a fairly long time to be dissipated, wax is less likely to form in the tank while the vehicle is being loaded or unloaded or during brief halts. Furthermore, except while initially igniting the heater, there is no drain on the battery. This system can also be adapted for keeping the engine running at optimum temperature, for example during operation at light load in extremely cold ambient temperatures. The disadvantages are high initial cost, long lengths of water

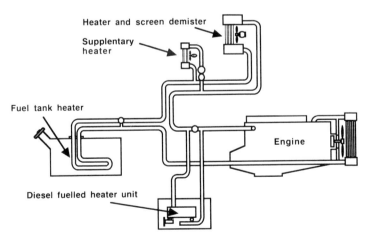

Fig. 4.13 Diagrammatic representation of the Eberspächer Arctic Fox system installed in conjunction with other heater units

and fuel pipes needing to be lagged if the additional cost of an electric heater in the filter and water separator is to be avoided.

4.10 Fuel feed pumps

The fuel feed pump for the diesel engine is the equivalent of a fuel lift pump for the gasoline engine. It draws the fuel from the tank and delivers it to the injection pump. Delivery of fuel both continuously and at a reasonable pressure is necessary because, if the extremely rapidly moving injection pumping plungers have to suck the fuel in, the lighter fractions of the fuel could form vapour bubbles in the pump and cavitation could occur. This of course would lead to uncontrolled variations in the rate of delivery of fuel to the cylinders, and therefore rough running and possibly even mechanical damage to the engine. The cavitation could also cause mechanical damage in the injection pump.

An external pump may be necessary for heavy commercial vehicles but unnecessary for cars, where the injection pump is relatively close to and not too much higher than the fuel tank, and where cost considerations override technical desirability. In any case, most cars have distributor type injection pumps, in which are transfer pumps delivering the fuel at a pressure adequate for moving the plungers during their inlet strokes. Some cars, however, also have hand-actuated pumps, as illustrated in Fig. 4.10, for priming and bleeding the system after, for example, filter changes and when the vehicle has been out of use for a long period. Injection pumps either are either automatically or manually bled, as described in Chapters 5 and 6.

For commercial vehicles, because of the variations in relative heights heights and distances between engine and tank, feed pump requirements vary enormously. The two extremes are the rear or under-floor engine bus with the fuel tank next to and at approximately the same level as the injection pump, and the truck with its engine mounted high forward of the cab and the tank low down a long distance to the rear.

Delivery pressures of between about 0.3 and 1 bar are generally considered to be adequate for preventing vapour formation on the suction side of in-line type injection pumps. The pressure is determined by the need to provide a good vent flow for both cooling the pump and reducing the differential heating effect from one end to the other of in-line pumps and, in a rotary or distributor pumps, to ensure that the supply of fuel is adequate for filling the plunger elements at high speeds. Feed pumps of the types that do not allow the delivery pressure to fall to zero between strokes are generally preferred. The delivery from the diaphragm pump tends to be less pulsating than that from other types, though there is little to choose between it and, for example, the AC electric pump described in Section 4. 13, which has remarkably few moving parts.

Diaphragms, however, represent a potential weakness, owing to their repeated flexure, though this type of pump has been in use for over a century and has been developed to an extremely high degree of reliability. Electrically actuated pumps are at an advantage relative to engine-driven mechanical pumps, in that their outputs are high under starting conditions, when a steady supply of fuel to the injection pump is especially desirable, and during idling when a pulsating supply of fuel could cause unstable running.

4.11 AC Unitac diaphragm type pumps

Diaphragm type fuel transfer pumps are similar to the lift pumps used for gasoline engines and described in Vol. 1 but, as can be seen from Fig. 4.14, of more substantial construction. The Unitac pump, Fig. 4.14(a), has a cam-and-lever actuation mechanism. As the outer end of the cam follower lever rides up the cam profile, its inner end, bearing down on a saddle at the lower end of the rod, pulls the diaphragm down, compressing its return spring. The pressure differential across the two valves above the diaphragm closes the delivery valve, on the left, and opens the inlet valve, on the right, thus allowing the fuel to enter the chamber above the diaphragm.

On the return stroke, because the inner end is forked to extend around the rod, it rises freely independently of the rod, leaving the diaphragm to be pushed upwards by its return spring. Again, the pressure differential actuates the valves, but this time, the inlet valve is closed and the delivery valve opened. The upward motion of the diaphragm, however, is limited to that needed to supply the demand for fuel from the engine, which of course varies considerably from idling to maximum power output. The function of the large chamber above the inlet valve is to ensure that the pump remains primed even if its feed pipe drains back to the tank.

As regards its basic principle of operation, the AC pump illustrated in Fig. 4.14(b) is similar to that described above, except in that it it designed for direct cam actuation, without a lever type follower, so the rod connected to the diaphragm is pushed up instead of pulled down by the cam mechanism. The return spring for the diaphragm is weaker than that for the push rod. Consequently, as the cam nose rotates past its top dead centre, the push-rod is pulled down by its return spring until the head formed at its upper end bears against the seating in the boss of the diaphragm carrier, pulling the diaphragm down with it. The inlet valve is opened by the pressure differential in the chamber above the diaphragm, which is therefore filled with fuel.

On the return stroke, the inlet valve closes and the delivery valve opens, allowing fuel to be delivered by the action of the diaphragm return spring, but only at the rate required to satisfy the demands of the engine. Therefore the cam rotates clear of the end of the push rod until, continuing round, it again lifts the diaphragm to fill the chamber above.

4.12 Bosch plunger type pumps

Bosch make two plunger type pumps, one single- and the other double-acting. The former suffers the disadvantage that in extreme conditions its delivery

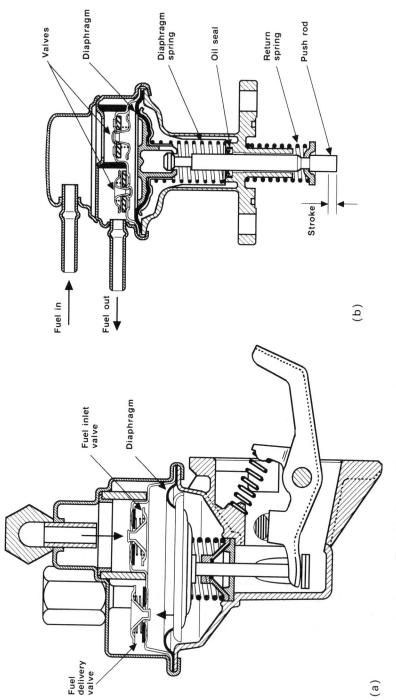

Fig. 4.14 Two AC lift pumps for diesel engines: at (a), the Unitac unit (a) is lever arm actuated, while that at (b) is directly actuated by an eccentric on the camshaft

pressure can fall to zero between pumping strokes; the other, however, does not. On the other hand, because the latter delivers continuously, it must have a pressure relief valve and either a bypass back to the inlet or a return line to the tank. Both are extremely compact, so much so that they can even be flange-mounted on the Bosch injection pumps, in which are situated the cams for actuating them.

From Fig. 4.15, it can be seen that the single-acting version comprises a cam-actuated piston in a cylinder flanked by non-return valves. As viewed in the illustration, the piston is lifted by its return spring, drawing fuel through the port on the right and a filter and non-return valve, into the pressure chamber. When the cam, acting on a roller-follower and push rod, forces it down again, the non-return valve on the inlet side closes and that over the transfer port on the other side opens, allowing the fuel to pass up into the chamber above the piston.

As the cam follower and the push rod rise, the piston is pushed upwards by its return spring but this time, because the pressure it generates closes the non-return valve over the transfer port, it displaces the fuel above it through the outlet port to the injection pump at the rate demanded by the engine. Since the piston lift varies with the demand for fuel, its stroke can vary from almost zero at idling to that required for sustaining maximum power output.

The double-acting version has no transfer valve: instead, it has two inlet valves and two delivery valves. With each stroke, fuel is drawn directly through the filter into the chamber on one side of the piston while, that on the other side is delivered to the injection pump which, as indicated previously, is why fuel in excess of the demand from the engine must be diverted through a pressure relief valve.

4.13 AC Alpha mechanical pump

At first sight, the AC Alpha pump, Fig. 4.16, looks like a positive displacement type unit, but it is not. Fuel enters at the top and leaves at the side. The piston on the left is push rod actuated, while the plunger on the right, which has a flexible top cover to prevent the entry of dirt and water, is used solely for manual priming.

When the cam lifts the piston, the inlet valve in the upper end of its cylinder closes and the fuel in the chamber above the piston passes through the transfer valve at its lower end into the chamber below. As the cam rotates past its top dead centre, the piston is moved down again by the two return springs above it, forcing fuel out through the delivery valve on the left, but at only the rate needed to satisfy the demands of the engine. Consequently, the push rod falls clear of the piston until the cam lifts it again. The spring that returns the piston is considerably stronger than that of the plunger on the right but, because the effective area of the latter is much smaller, it is not sucked downwards as the piston is lifted by its push rod.

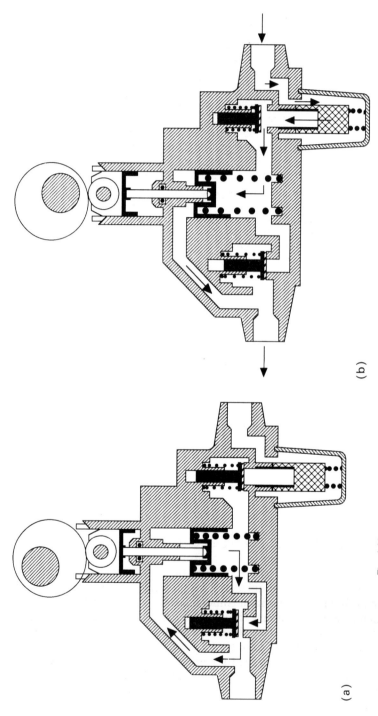

Fig. 4.15 Representation of the Bosch plunger type pump: at (a) in the fuel transfer and at (b) in the fuel delivery condition

TRANSFERRING THE FUEL FROM TANK TO ENGINE 59

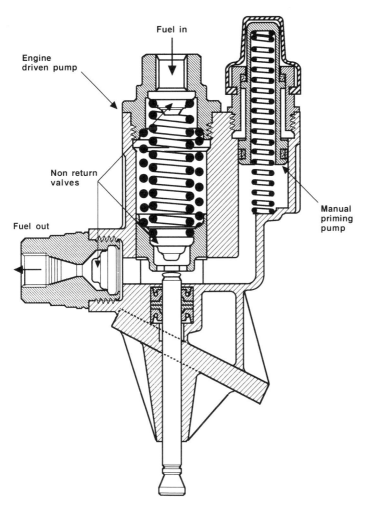

Fig. 4.16 An AC plunger type pump with, embodied on the right, a manual priming pump

4.14 AC Universal Electronic Solenoid fuel pump

The AC UES pump was originally designed in 1981 as a diesel fuel transfer pump, but has since been modified to render it suitable also for pumping gasoline and other fuels. Among the modifications are a more powerful coil, a special magnetic steel piston to improve its stroking and a Teflon piston ring. This ring seals particularly well, for rapid priming and handing hot fuel.

Among the advantages claimed by AC are few moving parts and none that wear out. It has been designed for operation at high temperatures, is quiet and, by virtue of its solid state circuits, reliable. Since the fuel is confined to a central tubular core, the coil assembly and electronics components

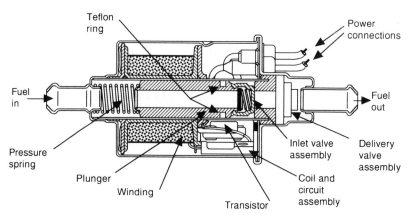

Fig. 4.17 The AC Universal Electronic solenoid type pump

around it are kept dry. A 12 V version will operate at voltages as high as 24 and as low as 8. Moreover, a 24 V version is available that will operate at voltages ranging from 14 to 24.

As can be seen from Fig. 4.17, When the coil is energised, it pulls the piston down against the resistance offered by the return spring. At the same time, the outlet valve in the delivery port closes, and the inlet valve in the piston crown opens and, as the piston moves downwards, fuel enters the chamber above it.

When the piston reaches the bottom of its stroke, the electronic system switches off the current to the coil and return stroke of the piston is powered by its spring, forcing the fuel above it up through the outlet valve. This sequence occurs many times per second. The quantity of fuel actually delivered is only that called for by the engine, so the piston stroke is rarely its maximum. To cater for reverse polarity voltage being inadvertently applied, the circuit is protected against burn out. The system is also protected against electro-magnetic interference and alternator dump loads.

Chapter 5
Injectors and in-line and unit injection pumps

Injection pump, nozzles and the pipes that interconnect them are interdependent, as regards the overall performance of the system. The pipes have been dealt with in the previous Chapter but, in this one, some information on the wave phenomena in the high pressure pipes will be added.

It would not be practicable in the space available to describe in detail all the most widely used injection pumps. Even so, the author has done his best to present as broad a spectrum as practicable. In any case, the aim is primarily at providing the reader with information on the basic principles. Since, technical illustrations and details of some of the injection equipment is virtually unobtainable in the UK, the author has had to confine both this chapter and that which follows to illustrating those principles by reference to a selection of the equipment actually in production at the time of writing, and for which illustrations are available. A few notes of historical interest have also been added.

Included are injectors and pumps produced by Stanadyne, unit injectors by Penske and Cummins, Lucas equipment and some of that produced by Robert Bosch GmbH, who offer the following range of injection pumps produced in very large quantities: their A Series is for 4- and 6-cylinder engines for cars, and the M, MW, P...S1 Series are for medium and the P...S7100 Series is for heavy commercial vehicles. These units are designed respectively for following maximum pressures at the nozzle of 400, 600, 900, 800 and 1200 bar, and for maximum power outputs per cylinder of 15, 25, 35, 45 and 70 kW.

In general, the demands on the injection system are exacting in the extreme. Indirect injection engines of only about 0.35 litres per cylinder are being produced. At full load, these consume only about 20 mm^3 of fuel per stroke and at idling only about 3 to 4 mm^3 per stroke. Moreover, even in a four-stroke engine, these quantities have to be injected precisely and consistently at a frequency up to about 2500 times per minute. The duration of injection must not exceed 30 to 35 degrees of crankshaft revolution, which is the equivalent of about a thousandth of a second. With direct injection, although the quantities to be metered are not so small, they do have to be delivered in even shorter times and against higher pressures in the combustion chamber.

5.1 Injectors

The main functions of the injector are: to define the pressure at which injection begins; to control the rate of injection for obtaining the required rate of increase in pressure and a combustion process that is both complete and does not generate harmful emissions; to inject the fuel into the combustion chamber in a manner such that all of it atomises, and mixes with and evaporates into the air as completely and uniformly as practicable within the time available; and, finally, to terminate injection instantaneously at the appropriate time, without any dribble or after-injection, leaving the combustion chamber completely sealed off from the fuel system.

Because of not only the need to aim the injection holes towards specific parts of the combustion chamber but also the extreme dimensional accuracy that needs to be preserved in service and the very close manufacturing tolerances, injectors for large engines are not usually screwed into the cylinder head as are sparking plugs. Instead, they are either flanged, so that they can be secured by bolts, or a claw or pair of claws are bolted down over either the flange or the body of the injector. Alternatively some sort of a saddle is fitted over them and bolted down. However, some injectors for small, indirect injection engines, for which compactness, simplicity and low cost are overriding considerations, do have M22 or M24 threads so that they can be screwed into the head.

In most injectors, a pre-compressed coil spring bears down on the upper end of a needle valve, and thus holds its lower end on its seat. If fuel leaking up the valve stem into the spring chamber were allowed to be trapped there, the injector would become hydraulically locked. Consequently, a pipe connection at or near the top of the injector body takes leakage back to the pump, filter or fuel tank.

The valve is lifted from its seat by the steeply rising pressure of the fuel acting on the cross sectional area equal to that of the bore of the guide less that of the valve seat. Immediately the valve has lifted, this area increases to that of the guide bore less the orifice. Rapid lift of the valve from its seat is assured because a considerably larger cross sectional area is suddenly exposed to the fluid pressure which, at the instant that it begins to lift, is generally between 50 and 280 bar. The initial lift rate is influenced also by the valve seat angle and width.

In general, the highest injection pressures are needed for turbocharged and the lowest for IDI engines. For direct injection, the lowest pressure at which the valve begins to lift is about 200 bar. High pressures, initially for valve lift-off and subsequently even higher for injection, ensure good distribution and penetration of the jet but, if fuel is deposited on the walls of the combustion chamber, emissions of HC tend to increase. For IDI engines, therefore, because of the small diameters of their combustion chambers and very high velocities of swirl, such very high pressures are not only unnecessary but also undesirable. As regards the capacity of the pump to supply the engine with fuel, the important criterion is the mean effective injection pressure (MEIP).

The high velocity of flow through the valve seating area, together with the tendency for evaporation to start, atomises the fuel. However, the rate at which it can be delivered and the pattern of the spray are determined by the effective cross sectional area, length and shape of the short passages past the valve head or through the hole or holes in the tip of the nozzle. In a pintle type nozzle, most of these parameters are to a major extent a function initially of the profile of the pin-like extension of its end, or pintle.

There are, however, designs in which poppet type valves are employed, opening downwards instead of upwards from their seats. In such cases, the fuel at high pressure can be delivered through the spring chamber to the nozzle, so there is no need for a leakage return pipe connection. With both types, gas pressure influences valve seating, but it tends to close the poppet and open those in the pintle and hole type injectors, which therefore require stiffer return springs.

5.2 The pintle type injector

There are three basic types of nozzle in common use: one is the pintle type already mentioned, the second is the hole and the third the two-stage, including Pintaux, type. The pintle type injector will be dealt with in this section. Because of the rapid air swirl in the combustion chambers of IDI engines, they do not need such a high energy jet as is characteristic of the hole type, so the pintle type is usually fitted. As can be seen from Fig. 5.1, the pintle valve, generally about 4–6 mm diameter, has a conical seating

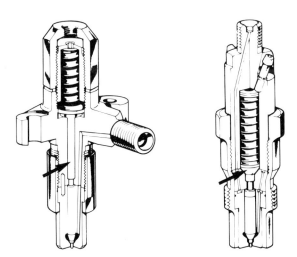

Fig. 5.1 Left, high-spring pintle type nozzle, in which the spring seat is on the upper end of a spindle (arrowed). Right, low-spring pintle type nozzle, in which the spring seat bears directly on the nozzle valve, thus reducing the inertia of the moving parts

face, below which its end is in the form of a pintle projecting into the hole. In general it lifts off its seat at a pressure of between 110 and 135 bar, dependent upon the engine.

The pintle does not rise fully out of the hole, and it is generally profiled so that, as it lifts, both the shape and volume of the spray are determined by the changing cross section of the annular clearance between it and the hole. An important requirement is that even the fuel initially injected is well atomised. From Chapter 7, it can be seen that the delivery characteristics of such a nozzle are conducive to high quality mixture preparation and quiet combustion.

Some pintles have parallel sides, others are waisted to form mushroom heads for spreading the jet as it issues from the hole, and some are simply tapered to produce a similar waisted effect but with a more abrupt change of section at the tip. Others have parallel flanks adjacent to the valve seat and then taper down, or are otherwise profiled, to increase the cross sectional area of the discharge orifice progressively as the valve lifts. This smooths the transition from the initially restricted to the full flow characteristic, and thus tends to reduce noise and improve driveability.

In the pintle type (Section 5.3) injector the area of the the clearance between the pintle and the hole is critical. If it is too small, the rate of lift of the needle valve is accelerated, tending towards noisy combustion. Moreover, at the end of injection, the rate of seating of the valve is reduced, again with detrimental effects but, in this case, mainly in respect of local carbon build-up, and smoke and unburnt hydrocarbon exhaust emissions. If, on the other hand, the hole is too large, the effect is mainly noisy combustion, due to too large a quantity of fuel being injected during the delay period. The termination of injection is not adversely affected: on the contrary, the more rapid closure of the valve can actually be an advantage as regards spray quality and absence of dribble.

Hole and needle size also influence the build-up of carbon. Some build-up, and therefore throttling of the fuel flow, is inevitable, but its rate can be reduced by appropriate design measures. The smaller the annular gap between pin and hole the greater is the self-cleaning effect and the slower and more uniform therefore is the carbon build-up. Uniformity of discharge orifice shape and section is of course important as regards production of the optimum shape of the spray and quality of mixture preparation.

For all injectors, it is important to ensure that the needle valve both seats and seals properly at the end of injection. Pressure waves in the fuel and vibrations in the needle seating spring and the mechanism in general can contribute to or even cause unsatisfactory sealing. Failure to seal properly, especially in highly rated turbocharged engines, may result in blow-back of the combustion gases and consequent instability in the hydraulic system. Influencing factors include seating area, hole area, needle diameter, and spring rate and pre-compression. Precision and close manufacturing tolerances are of course important too. Stems and guides used to be manually lapped as pairs, the tolerances being of the order of 0.0025 mm. Now, however, the needles are either selectively assembled into the guides or they are match ground to pre-ground and honed bores in the nozzle. Consequently valves and guides or bodies must never be interchanged.

The rate of carbon build-up is dependent also of course on temperature. If the temperature of the nozzle rises above about 220 deg C, some sort of shielding may be necessary to reduce the rate of heat transfer to it. The shield, which may take the form of a short sleeve around the tip of the nozzle, or simply a counterbore in the cylinder head casting, conducts the heat away to the head. If shielding alone proves to be inadequate, it may be necessary to provide nozzle cooling.

5.3 The hole type injector

The hole type injector is used mainly in direct injection engines. It may have one hole, but more often has two or more, and is available in several forms. For ease of drilling, the axes of the holes should be perpendicular to the external surface of the nozzle tip. If the injector is inserted vertically into the centre of the roof of combustion chamber in a four-valve head, the holes may be symmetrically arranged around the tip, as in Fig. 5.2. On the other hand, if it is inclined and off centre in a two-valve head, the holes are more commonly asymmetrically arranged, Fig. 5.3, to distribute the spray symmetrically in the combustion chamber. In this case, accurate rotational location of the injectors is of course essential.

Hole type injectors also have alternative seating arrangements. One, variously termed the *VCO, valve covered orifice* or, sometimes, the *seat hole* type nozzle, is illustrated in Fig. 5.4(a), while a two-hole nozzle of the *conical blind hole* type is illustrated in Fig. 5.4(b), and a *cylindrical blind hole* type in Fig. 5.4(c). In general, needle lift is initiated at a pressure of between 150

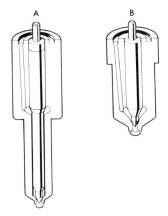

Fig. 5.2 Lucas produce both long stem (A) and short stem (B) type multi-hole nozzles for direct injection engines. The long stem type is for engines in which adequate cooling cannot be arranged for short stem nozzles

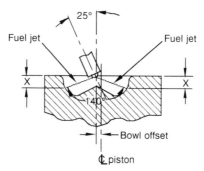

Fig. 5.3 Where the tip of the nozzle is coincident with the axis of the combustion chamber, and the jets are directed at equal angles to each side of that axis, the jets strike the combustion chamber wall at equal distances X below the piston crown

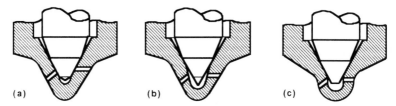

Fig. 5.4 Alternative valve seating arrangements for hole type nozzles: (a) valve covered, or seat hole, type nozzle; (b) conical blind hole type; (c) cylindrical blind hole type. Where the holes are assymetrical, as in this illustration, provision has to be made for accurate directional location relative to the combustion chamber

and 250 bar. This is higher than that for the pintle type, because the higher peak pressures experienced in direct injection combustion chambers at the time of valve closure could otherwise cause blow-by.

The *sac volume* is that of the space between the end of the sac and the tip of the needle when it is on its seat. Keeping the sac volume of the nozzle to a minimum is important. After injection has terminated, the high temperature of combustion may evaporate fuel remaining in this space, and force it out into the combustion chamber where, entering late in the combustion process when little air remains to burn it completely, it tends to generate black smoke and unburned hydrocarbons in the exhaust gases.

Therefore, with the ever increasing stringency of emissions regulations, the trend has been towards reducing the sac volume. This can be reduced by shortening its length, which may necessitate truncating the conical end of the needle. On the other hand, truncation is sometimes employed simply as a device to facilitate the flow of the fuel to the hole. In some instances, the end is of spherical form, so that it can fit even more closely in the sac. Truncated conical and spherical tips, which are options offered by Stanadyne on their Pencil Injector, are compared in Fig. 5.5.

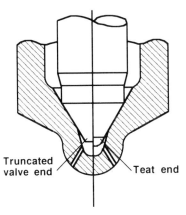

Fig. 5.5 Stanadyne have developed both truncated and spherical tips for their Pencil injector

5.4 High and low spring injectors

Typical high and low spring injectors produced by Lucas are illustrated respectively in Fig. 5.1(a) and (b). Originally high spring injectors such as that at (a) were virtually universal, but later the low spring type (b) came into use because of the smaller mass of its reciprocating parts.

In the Lucas LRC low spring injector, two back-leak connectors fitted to the upper end of the spring chamber return leakage past the needle to the filter or tank. For adjustment to the pre-loading of the spring, the injector has to be dismantled so that an appropriate shim can be interted beneath the springs. Leakage up the needle valve guide passes up through a duct into the spring chamber, and thence away through the two previously mentioned connectors.

Although low spring injectors were originally developed for turbocharged engines, they are now more widely used than the high spring type, even on the naturally aspirated power units. This is because, owing to the lower inertia of the needle valve and spring assembly, they close more rapidly and therefore are more readily adapted for meeting emissions regulations. Needle valve lift characteristics are compared in Fig. 5.6.

5.5 Two-stage injection nozzles

Many years ago, the Pintaux nozzle, Fig. 5.7(a), was developed jointly by CAV and Ricardo, for application to the Ricardo Comet IDI combustion chamber. The aim was at making provision for cold starting and, in that such nozzles are better than the single stage type, they were a significant improvement. Since then, however, most of these engines have heater plugs, which are even better as aids to cold starting, so very few engines now have

68 INJECTORS AND IN-LINE AND UNIT INJECTION PUMPS

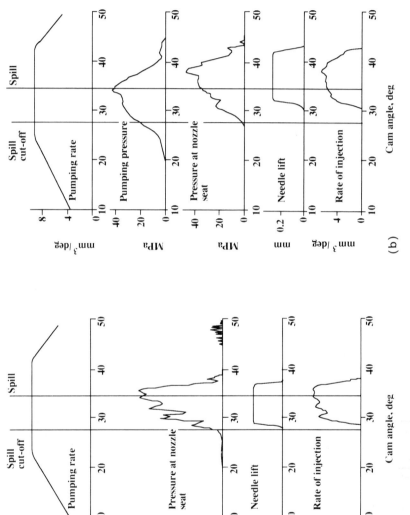

Fig. 5.6 Two sets of injection diagrams, for (a) a unit injector and (b) a pump-pipe-nozzle system

INJECTORS AND IN-LINE AND UNIT INJECTION PUMPS

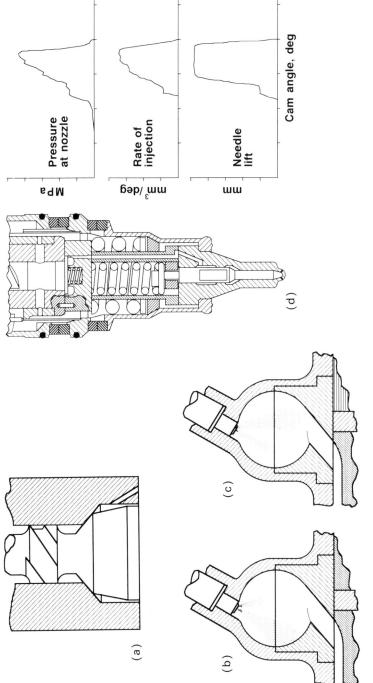

Fig. 5.7 The Pintaux nozzle (a) is little used now because its place has been taken by more modern two stage nozzles for indirect injection engines. At (b) it is shown operating at full lift and at (c) during the pilot injection phase, just as the valve is beginning to lift. A typical dual spring two-stage injection nozzle and an associated injection characteristic diagram is shown at (d)

this type of injector. The term Pintaux is an abbreviated combination of the words pintle and auxiliary jet.

With the pintle type nozzle described in Section 5.2, the spray is normally coaxial with the injector. For all round engine performance it is best directed downstream into the air swirl, towards the side of the combustion chamber remote from its throat. However, for cold starting, it is better injected into the centre of the vortex, where the air is remote from the cooling effect of the walls of the combustion chamber and therefore warmest.

To satisfy both of these requirements, the Pintaux nozzle has two holes, one axial and a smaller one directed obliquely, as shown in Fig. 5.7(b) and (c). Its pintle is of a shape such that it partially blocks the axial hole until the needle is lifted well clear of its seat. Initially, therefore, injection occurs mainly through the smaller oblique hole, which directs it to the centre of the combustion chamber. Subsequently, when the pin lifts out of the axial hole, about 90% of the fuel is injected through it in the downstream direction, leaving only 10% passing through the smaller diameter oblique hole.

An alternative method of producing a two-stage injection effect, but in a direct injection engine, has been employed by Perkins in their 2.0 litre diesel unit for the Freight Rover. In this instance, both the initial and the follow up jet are directed along the same axis. Two springs of different rates are installed in series in a Lucas injector, the needle valve being lifted initially against only the low rate spring. Then, as the pump pressure builds up, the second stage spring is compressed and fuel injected at high pressure. As engine speed increases, so also does the rate of build-up of pressure during the injection process. Therefore, the ratio of initial to main fuel delivery rate decreases with engine speed, thus providing optimum noise-reduction at low speeds with neither a smoke nor a power penalty at higher speeds and loads. A typical two-stage injection nozzle is illustrated in Fig. 5.7(d).

Without a two-rate spring, some method of generating two pressure stages during injection would have to be devised. In this case, difficulty would be experienced in avoiding poor atomization at the low rate. Moreover, to deliver the same quantities as with single stage injection, the whole injection process would have to be spread over a longer period. This in turn, would require the maximum rate to be reduced and therefore, even at the fairly high rates towards the end of injection, atomization might not be so good as with the equivalent single-stage system.

5.6 The Stanadyne Pencil injector

The Pencil Nozzle, Fig. 5.8, was invented in the late 1950s by Vernon Roosa, who also introduced the first Stanadyne distributor type pump. The name Pencil Nozzle was given to it because the spring chamber and nozzle body had been reduced from the then typical overall diameter of 21–25 mm to 9.5 mm.

Subsequently, the Slim Tip version of the Pencil Nozzle, with a nozzle extension of only 5.4 mm diameter, was introduced. It went into production

INJECTORS AND IN-LINE AND UNIT INJECTION PUMPS

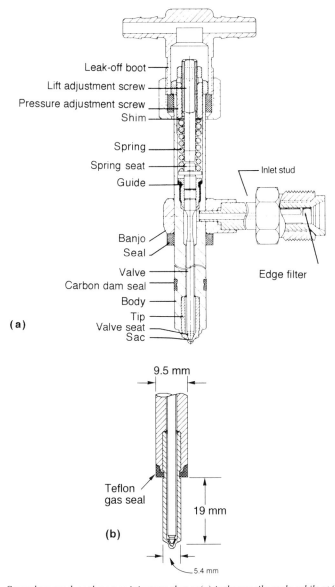

Fig. 5.8 Stanadyne produce these two injectors: that at (a) is the pencil nozzle, while at (b) is shown how the Slim Tip version is produced simply by modifying the lower end of the Pencil type

in 1984 and, in 1985, was fitted to the Ford 2.5 litre high speed direct injection engine for the Transit van. Both versions are of the hole type. A comprehensive description of the development of both is given by Lewis E. Tolan, in a paper entitled 'The Pencil Nozzle, Its Improvements, Advantages, and

Future', presented at the *Third Congress of the International Association for Vehicle Design*, in Geneva in 1986.

Among the advantages claimed for the Pencil Nozzle are the following:

((1) With a shank diameter of only 9.5 mm, provision in the cylinder head casting for cooling not only the injector but also the components around it is simplified
 (2) It offers the engine designer more flexibility as regards layout of valves, porting and combustion chamber in the head
 (3) It is secured by a three-leaf spring-claw, Fig, 5.9, one end of which is bolted to the head casting, and the free end bears down on the banjo connection for the fuel inlet
 (4) By siting the spring remotely from the combustion chamber, risk of its relaxation due to high temperatures is reduced.
 (5) The weight of the injector is only 140 g.

The Slim Tip version, Fig.5.8(b), of the Pencil nozzle, Fig. 5.8(a), differs only in that there is no carbon dam seal on the shank of the nozzle and the slim tip, crimped into the body, has a Teflon gas seal around the shoulder between the two parts. The latter serves as a carbon dam and replaces the more widely used copper seal, thus obviating the need for heavy clamping loads. Consequently, potentially high local stress concentrations and bowing of the slender body, are avoided and both manufacturing tolerances and differential thermal expansion are more easily accommodated.

This version has two advantages in addition to those already mentioned in the context of the pencil nozzle. Both arise because the diameter of the

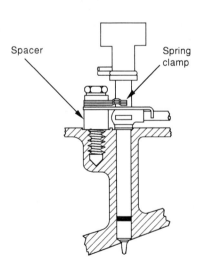

Fig. 5.9   A spring clamp is used to hold down the Stanadyne injectors. The thickness of the spacer should be such that its top face is in the same plane as the top face of the banjo, with a tolerance of 0.5 mm. This gives a clamping load of 227 to 544 kg

slim tip is only 5.4 mm. First, it absorbs less heat from the burning gases and, secondly, it has potential for installation even closer to the centre of the combustion chamber than the original version.

From Fig. 5.8(a), it can be seen that fuel from the injection pump is delivered to a banjo connection approximately mid-way between the upper and lower ends of the pencil injector body. Above it is the guide and spring assembly, while below is the valve and nozzle assembly. The inlet connection houses an edge type filter. Adjustment to the compression in the spring is effected by screwing the sleeve that forms its upper seat up or down in the end of the body. The height of the lift stop is adjusted simply by screwing it inwardly or outwardly in the sleeve. Two lock nuts are used, to secure each adjustment independently of the other.

A glass-filled seal beneath the banjo prevents water and debris from entering the housing bore in the cylinder head and controls the projection of the tip into the combustion chamber. Lower down, in a W groove to allow for spreading of the material into it under radial compression, is a Teflon carbon dam, or seal, which is an interference fit in the bore. Just above the point of entry of the fuel into the body is the valve stem guide.

Two grooves in the periphery of the valve stem form reservoirs to receive the fuel leaking back and then spread it around the bore of the guide, for lubrication.

Beneath the guide, the stem is for three reasons waisted or, in American terminology, necked. First, it facilitates straightening following induction tempering during manufacture. Secondly, it behaves rather like a universal joint for automatically adjusting for any misalignment between guide and seat. Thirdly, it provides an easy flow path for the fuel as it turns down from the inlet connection towards the valve tip.

The diameter of the valve shank is 0.085 in (2.16 mm) and that of the bore in the body 0.105 in (2.67 mm). Thus the cross sectional area of the clearance between the two is approximately equivalent to that of the bore of commonly used high pressure tubing for delivering the fuel to the banjo connection. It has never been found necessary to match the two areas any more closely.

5.7 Types of fuel injection pump

Various pump configurations and principles of pumping have been applied over the years. Now however, for automotive applications, the three most commonly used types of fuel injection pump are:

(1) The in-line pump, which has a row of cam-actuated pumping elements, one for each cylinder of the engine, above a camshaft. This type is most commonly fitted to diesel engines for heavy commercial vehicles.
(2) The distributor type pump, in which several of the engine cylinders are served by one, two or more plungers, from which the fuel is directed to each cylinder by a distributor device. This type predominates on cars and light commercial vehicles.

(3) The unit injector, which is a combined pump and injector unit, one of which serves each cylinder. All are actuated by the engine camshaft. The principal advantage of this type is that fuel at the very hgh pressure needed for injection is contained in very short ducts in the injector itself, instead of having to be delivered through long and relatively vulnerable pipes. However, unlike the more compact injectors used with the other types of pump, it is not easy to accommodate on the cylinder heads of the smaller engines.

Other systems are under development, though at the time of writing, none has gone into quantity production. One has a separate injection pump for each cylinder, with a very short pipe connection, or possibly ducts in the cylinder head casting, between it and the injector. This largely overcomes the problems associated with long high pressure pipe lines, as explained in Sections 5.9 and 5.10. It also can be easier to accommodate on the cylinder head, because a compact standard injector can be installed where it does not overcrowd the region of the valves, and the pumping element can be mounted closer to and directly actuated by the camshaft. A major disadvantage is its relatively high cost.

Another system, which is currently being pioneered by Caterpiller, has a hydraulic accumulator serving a common rail high pressure supply system to the injectors. To provide very high injection pressures, a hydraulic pressure multiplying device is interposed between the pump and injectors. Among the problems appears to be the time lag between delivery from the pump and the supply of fuel through the various devices to the injectors. Few details have been made public regarding any of these new systems, so not much more can be said about them here.

In general, unit injection seems likely to become the system most widely used in future to combat exhaust emissions. It is is less costly in terms of hardare than the other systems, but has the disadvantage that it cannot be applied to existing engines without major redesign and development, and the installation of new production equipment. Consequently, engine manufacturers are having to wait until the capital investment entailed can be justified.

5.8 In-line injection pumps

The functions of the injection pump are threefold. First, it must supply the fuel at precisely the right point in the diesel cycle for the pressure build-up resulting from ignition to occur at and after tdc. Until the advent of stringent emissions regulations, the timing of the start of injection was generally fixed. Now however, it may have to be varied with engine speed and load.

Secondly, since the chemical energy content of the fuel injected primarily determines the power output, the pump must meter the quantity delivered per cycle to match the torque-speed relationship demanded by the driver. Moreover, the fuelling must be balanced (fuel supplied equally per injection to each cylinder).

Thirdly, the pressure energy in the fuel must be very high, to enable the injector effectively to atomise, mix and evaporate it in the air in the extremely short interval of time available in the cycle. Moreover, to obtain a high rate of delivery in the shortest possible time in a high speed engine, the delivery characteristic should be in principle a rectangular pulse, giving an extremely steep rise and termination of pressure with, in between, a flat top. In practice, however, it may have to be shaped both to meet emissions requirements and to keep noise down to acceptable levels. The way in which the delivery characteristic is timed in relation to the compression and power strokes can be seen from Fig. 5.6.

Maximum injection pressures, as distinct from the valve opening pressures quoted in Section 5.1, depend upon the type of combustion chamber and whether the principle of fuel finding air or air finding fuel, as explained in Chapter 7, is adopted. Indirect injection engines, with their very high swirl rates, call for maximum injection pressures of about 300–400 bar. For direct injection engines, the pressures have hitherto been of the order of 450–850 bar though, to satisfy the continually tightening emissions regulations, pressures of up to 1200 bar are already established, and the trend is still upwards. For engines without significant air swirl, pressures of over 1000 bar have always been essential.

Basically, high injection rates are generated by high pressures in association with small holes. The larger the quantity of fuel that can be injected per cylinder per stroke in a given time, the greater is the potential for increasing either the torque, or the speed and therefore power output, of the engine. There is an upper limit, however, because nothing can be gained by injecting at a rate faster than that at which the fuel can be mixed with the air and burnt. In any case, the outcome would be black smoke in the exhaust. Moreover, the rate of flow through the injectors cannot exceed, or even closely approach, the velocity of sound in the fuel, which is about 1250–1500 m/sec, depending mainly on the temperature of the fuel, but partly also on the geometry of the pipeline (the comments, in Chapter 11, regarding the factors affecting the velocity of sound in induction pipes, are relevant here, though the velocities in fuel pipes are different).

5.9 Fuel delivery characteristics and injection lag

As mentioned previously, the delivery characteristics of the pump and injector nozzle combination must be such that maximum pressure is attained in the shortest practicable time. Moreover, the cut-off of flow through the nozzle must be sharp, so that there is no dribble, and it is also important to avoid after-injection due to the re-opening of the needle valve by pressure waves reflected back to it from the pump.

Dribble or after-injection can lead to a build-up of carbon on the nozzle, which eventually can even block the nozzle completely. If fuel does dribble through the injector hole or holes, it comes out in large droplets, which is why it causes local carbon build-up, and leading ultimately to interference with, or even total blocking of the main flow. Moreover, the fuel thus delivered is burnt so late in the cycle that the efficiency of conversion of its energy into heat is low. All this adversely affects performance and leads to smoke and other undesirable emissions.

Several measures can be taken to contribute to the elimination of these phenomena. One is the use of a spill port, Section 5.11, to release the pressure from the pumping element virtually instantaneously. Rapid seating and effective sealing of the needle valve in the nose of the injector nozzle is of course essential. Also the incorporation of unloading collars in the delivery valves on the pumps, Section 5.12, can generate a large negative pressure wave immediately prior to their closure. Another measure is to install a snubber valve, Section 5.10, to damp the pressure waves.

Basically, rapid build up and collapse of delivery pressure is a function of cam design. In this context, a point to be considered is that, at high speeds, the inertia of the pumping plunger tends to cause it to over-shoot beyond its top dead centre. This is liable to cause mechanical damage in the event of over-speeding of the engine. It can be avoided by profiling the cam so that the upward acceleration of the plunger is progressively reduced just as it approaches tdc.

There is a time lag between generation of the pressure at the pump and the delivery of fuel into the cylinder, which must be taken into account This, termed *injection lag*, is attributable to two effects. One is the time taken for the wave due to the compression of the fuel by the pump to travel the length of the delivery pipe. The second is the time taken by the injector valve to lift off its seat after the pressure wave has reached it. Thirdly, waves generated by the pump, and the seating and unseating of the delivery valve and injector needle are reflected back and forth along the pipe. Other factors causing pressure waves in the fuel flow are camshafts or drives which vibrate torsionally owing to inadequate rigidity, or backlash in drives.

Some idea of the compressibility of fuel oil can be gained from Fig. 5.10(a), in which the line inclined at slightly less than 45 deg represents the compressibility and the curve the bulk modulus, which is defined as the reduction in the bulk of the fuel per unit increase in pressure. Bulk modulus varies with temperature, Fig. 5.10(b). Incidentally, waves may be generated too by the radial (ballooning) deflections of the pipe between the pump and injector.

The fuel compression effect is a function of the length of the pipe line which, for this reason and to keep is natural frequencies of vibration as high as possible, should be as short as practicable. Injection lag due to the initial fuel compression wave, expressed in terms of degrees of rotation of the pump shaft, increases with engine speed. Provided the pipe line is stiff enough, the ballooning effect is very small, generally to the point of being negligible.

From Fig. 5.6(b), reproduced from SAE Paper 851458, Fuel Injection in Automotive Engines, by Dr P.E. Glikin it can be seen that, owing to the speed of motion of the plunger and the progressively increasing restriction to flow offered by the closing inlet port, the pressure in the pump begins to rise slowly over an initial period of about 2.5 deg of camshaft rotation after

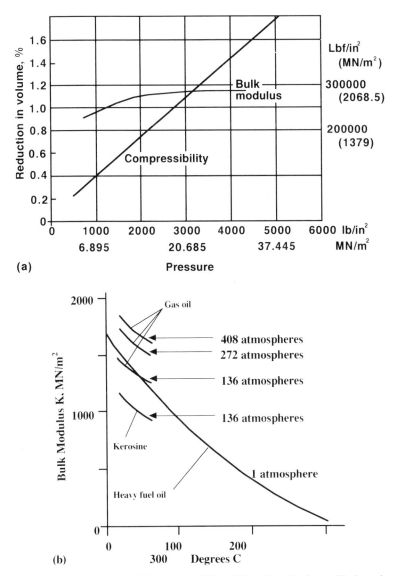

Fig. 5.10 The curve at (a) is of the compressibility of fuel oil, while that at (b) shows how the bulk modulus of various fuels vary with temperature

the inlet port begins to close. Over the next 2.5 deg it progressively increases up to the cut-off point, before continuing much more steeply to the point at which it levels off to form the plateau. For similar reasons, it finally collapses over a period of about 5–6 deg beyond the spill point. With unit injection, the pressure rises earlier to its higher maximum and also collapses more rapidly. Consequently, for any given quantity of fuel to be delivered, less time is required for injection, which therefore can start earlier without risk of late burning.

The illustration also demonstrates the lag between pressure changes at the pump and the nozzle seat owing, as previously mentioned, to the time needed for the pressure wave to travel along the pipe. When the pressure wave arrives at the nozzle, there is of course a further delay until the pressure there has built up to a level adequate to unseat its valve. Similarly, injection ceases only when the wave originated by the opening of the spill port has reached the nozzle. All these effects become increasingly pronounced as engine speed rises. The ratio of mean pumping delivery rate to mean injection rate, sometimes called the *spread-over ratio*, can rise from 1:1 at low speeds to about 2 or 3:1 at maximum speed.

5.10 Pressure waves and cavitation

The effect on injection rate and pressure at the nozzle, of the previously mentioned waves reflected back and forth in the pipelines between the pump and injector nozzles, can be seen as high frequency waves superimposed on the curves in Fig. 5.6. Indeed, these reflected waves, sometimes referred to as the *hydraulic hammer* effect, can increase the pressure at the nozzle relative to that at the pump by as much as about 30%. Moreover, they can cause cavitation erosion of the valves, pipes, and their end-fittings. The shorter and stiffer the pipes, the higher are the frequencies of the pressure waves and the smaller their amplitudes. Events that can generate them include the opening and closing of the delivery and injector needle valves.

To prevent cavitation due to the pressure wave initiated by the sudden closure of the needle valve in the injector nozzle at the end of the effective stroke of the pump plunger, a return flow restrictor, or snubber valve, such as that in Fig. 5.11, may be installed downstream of the delivery valve assembly. As the delivery valve is lifted off its seat by the pressure generated by the pump, so also is the flow restrictor valve above it. Consequently, the fuel passes totally unhindered through both. When the effective stroke terminates, the restrictor valve return-spring causes it instantly to seat again. However, it has a hole in its centre, through which the reflected wave can pass, but its diameter is such as to damp wave motion. Another type of flow restrictor valve, in which the pressure wave unseats a ball valve, can be seen in Fig. 5.25. Unloading collars such as that described in Section 5.12, para 6, are also widely used.

5.11 In-line pumps — fuel metering

Essentially all in-line diesel injection pumps embody one or more cylinders, termed *barrels*, in which a reciprocating plunger generates the very high pressures required. Various types of valve have been used, with the aim of

INJECTORS AND IN-LINE AND UNIT INJECTION PUMPS 79

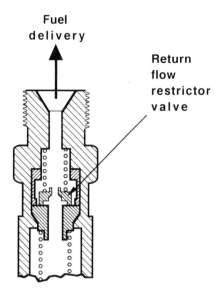

Fig. 5.11 Diagrammatic representation of the Bosch snubber valve

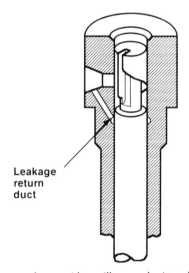

Fig. 5.12 An injection pump plunger with a spill groove having a helical edge

clearly defining the start and end of injection, but the only kind suitable for operation at the speeds demanded for modern engines are those in the form of ports that are covered and uncovered by the end of the plunger and the edge of the *spill groove*, Fig. 5.12.

As the plunger is jerked upwards by the cam, its upper end covers the fuel *inlet port*, at which point pressure in the fuel above it begins to build up, initially slowly and then extremely rapidly, as described in Section 5.9. Then, to terminate injection, this pressure is suddenly released by the uncovering of the *spill port* as the edge of the spill groove lower down the plunger passes over it. The spill groove is generally straight but inclined so that, by rotating the plunger, the timing of the *spill point*, and thus the quantity of fuel injected, can be controlled to match the torque demanded by the driver of the vehicle.

For rotating the plungers, some early injection pumps had gear teeth machined on their control rods, to mesh with pinions near the lower ends of the plungers, forming a rack-and-pinion mechanism. Consequently, the term *rack* came to be, and has tended to continue to be, used to describe the control rod, even though the plungers may be rotated by levers, as described in Section 5.12.

Note that the timing of the cut-off of the inlet port is constant, though the start of injection can be varied by introducing a mechanism for rotating the camshaft a few degrees relative to the drive shaft. Alternative plunger designs are described in detail in Section 5.14. In unit injectors, metering the fuel is effected as described above, though in distributor type pumps it is entirely different, as explained in Chapter 6.

In all types, keeping to a minimum the leakage between the plungers and their barrels is of prime importance. The barrels are ground and lapped, and the plungers either selectively assembled into them or match ground and lapped to them. Consequently, the plungers and barrels must be kept in pairs in service, to ensure that their components are not inadvertently interchanged.

5.12 Lucas Minimec, Majormec and Maximec pumps

This is a range of three in-line pumps, all basically of similar design, all of which are driven from the end remote from the governor and differ mainly in size. The Minimec is produced with camshaft lifts of 7 and 9 mm, while the Majormec, available for only six- and eight-cylinder engines, has lifts of 10 and 11 mm. Maximec is an uprated and strengthened long-life unit and is currently manufactured for only twelve-cylinder engines for military and railway locomotive applications.

Since all operate on the same basic principle, only the Minimec, Fig. 5.13, will be described here. It is available in three-, four-, six-, and eight-cylinder forms. Although it is normally offered with the pump and governor in a one-piece housing, it can be supplied with a two-piece housing, the two being bolted together. The casting that houses the camshaft, tappets and control linkage is of light alloy, but the pumping elements are carried in a one-piece steel head. This arrangement has a number of advantages. First, it is very stiff and is relatively easy to seal effectively; secondly, variations in clearances due to differential thermal expansion are obviated; thirdly, it enables the pumping elements to be more closely spaced than if they had been in light alloy housings. This not only makes the pump more compact but also shorter, so the camshaft is inherently stiffer.

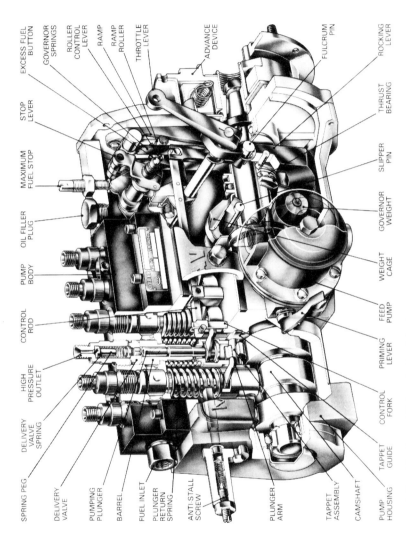

Fig. 5.13 The Lucas Minimec in-line injection pump

Stiffness is important, so a large diameter alloy steel camshaft is employed. It is carried in taper roller bearings in steel rings cast into the light alloy pump housing and governor cover. Above it, roller tappets reciprocate in bores in the aluminium housing. In the 9 mm lift version, flats on their peripheries register against steel keeper plates, one between each pair, to prevent them from rotating. In all the other versions (Maximec as well as the 7 mm lift Minimec) T-pieces between the tappets guide the ends of the rollers. Incidentally, In the Maximec, the camshaft bearings are in cast iron housings.

The foot of each plunger bears on a plate seated in the end of the tappet. These plates are manufactured in a range of thicknesses, so that they can be selectively assembled to ensure that the tappet rides on the appropriate section of the cam profile and to set accurately the phase relationship between the pumping elements. This is for compensating for manufacturing tolerances.

Each tappet and plunger assembly is compelled by a coil spring to follow the cam and base circle profiles. The camshaft and tappets are lubricated by engine oil piped to the unit, where its flow rate is metered by passing it through the small clearance between one of the tappets and its bore. Some earlier pumps were filled with oil on assembly but, owing to dilution with fuel oil leakage, their sumps had to be drained and refilled periodically.

Each plunger barrel is retained by the delivery valve holder, screwed down from above to clamp it against a shoulder in the steel housing, and the delivery valve seat is between the delivery valve holder and the end of the barrel. As can be seen from Fig. 5.14, a coil spring holds the fluted delivery valve down on its seat in the upper end of the valve guide inside the holder. Around the valve stem, between the flutes and the valve seat, is what is termed an *unloading collar*. When the plunger has risen to the point at which delivery occurs, the valve lifts off its seat. Just before the valve returns to its seat, the unloading collar enters the valve guide, in which it is a close fit. It therefore draws fuel back into the upper end of the guide, thus reducing the pressure in the delivery pipe and avoiding both after-injection and dribbling through the injector nozzle.

A fuel delivery gallery is drilled longitudinally in the steel head to serve the inlet ports drilled radially in the barrels. Fuel from the feed pump is delivered through a pipe connection that can be fitted to either end of the injection pump, to suit the engine installation. The feed pump can be fitted either to the pump and actuated by a cam on its camshaft, or mounted on the crankcase and driven by the engine camshaft.

As indicated in Section 5.11, the quantity of fuel supplied, and thus the engine torque, is regulated by rotating simultaneously the plungers serving all the cylinders, by means of a lever on the lower end of each. The outer end of the lever registers between the arms of a fork in which it slides vertically when the plunger is actuated. These forks are carried on a square section rod that slides axially in bearings in the ends of the housing.

These forks are assembled on to the rod in a jig and then nipped in position by set-screws, arranged in a manner similar to that illustrated in Fig. 5.13. The screws are finally tightened during pump calibration, when the pumping

INJECTORS AND IN-LINE AND UNIT INJECTION PUMPS

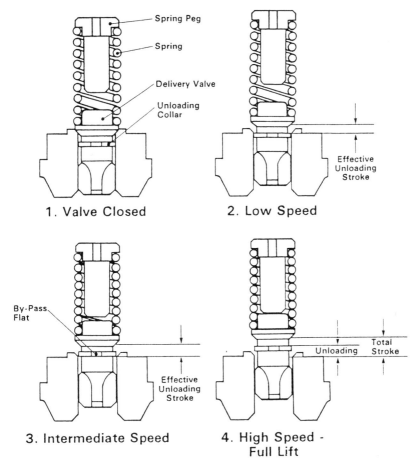

Fig. 5.14 The four stages of operation of the Lucas bypass type delivery valve, which has a flat on the unloading collar to allow a small quantity of fuel to bypass it. By virtue of the chamfered mouth of the valve port, the unloading operation is progressively terminated. Thus potential sources of severe shock wave generation are reduced

elements are balanced to give equal delivery to all the cylinders. Access to the screws is gained through a cover plate bolted over an aperture in the housing. The engine speed is of course regulated by the governor, the output lever of which is attached to one end of the rod.

5.13 Pumping element operation

The modes of operation of pumping elements in general are, with only slight variations, similar. In the Minimec, a straight gash is machined in the periphery near the upper end of each plunger, which is counterbored, in a

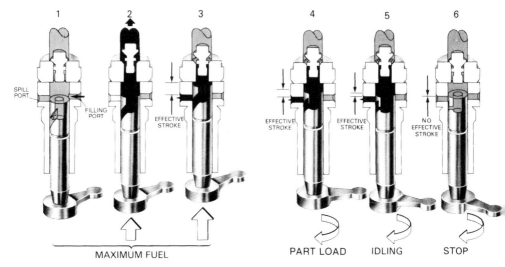

Fig. 5.15 Left, three diagrams showing how reciprocation of the plungers delivers the fuel to the cylinders. Right, these three diagrams show how rotation of the plungers varies the spill timing to regulate the rate fuelling of the engine

manner similar to that shown in Fig. 5.15. The axis of this gash is inclined relative to that of the plunger. This is much easier to machine than most of the other arrangements to be described later. Radial holes are drilled from the base of the gash into the counterbore.

As the plunger is moved upwards by the cam, its upper edge covers the inlet port, so the fuel above is then forced out through the delivery valve, as described in Section 5.12. Further movement of the plunger causes the gash to uncover the spill port, at which point the delivery pressure collapses suddenly because the fuel remaining above the plunger is returned through the counterbore, gash and spill port to the delivery gallery. When the plunger retracts, the inlet port is uncovered by the gash, allowing fuel to pass through it and on through the radial holes and counterbore to the space above the plunger, in readiness for the next delivery stroke.

The instant the upper end of the plunger covers the inlet port or ports is called the *cut-off point*, and that at which the edge of the gash uncovers the spill port is termed the *spill point*. Whereas the cut-off point is set to occur at a specific plunger lift to ensure that all cylinders receive fuel at a similar rate, and therefore is constant, the spill point is varied by the driver, through the medium of his accelerator pedal and the governor, to regulate the volume of fuel supplied and thus the power output. Both points can be specified in terms of either crank angle or plunger lift.

The plunger, reciprocating in the barrel, is lubricated by fuel, which must not issue from the lower end of this assembly because, otherwise, it would dilute the lubricating oil. To trap it, therefore, a peripheral groove is machined

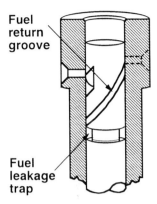

Fig. 5.16 The lower peripheral groove traps fuel that leaks down the plunger during the injection stroke, and returns it up the spiral groove when it passes the spill port

near the lower end of the plunger whence, during the delivery stroke, it is returned upwards through a spiral groove to the inlet port, Fig. 5.16.

The various phases of plunger movement are expressed by the following terms. From the point at which the plunger closes the inlet port (the *cut-off point*), the *pressure stroke* starts. The period during which the unloading collar is either rising out of or lowering into the upper end of the valve guide is termed the *retraction stroke*. This is immediately followed by the *effective stroke*, which terminates immediately the spill port is uncovered by the spill edge. The remainder of the stroke up to top dead centre is termed the *residual stroke*.

As the plunger retraces its path, fuel first passes into the space above it until the spill port is again closed, after which a vacuum is created in that space until the inlet port is uncovered and fuel can resume flowing in ready for the next pressure stroke. The length of the stroke between the closing of the spill port and the opening of the inlet port increases with the load on the engine: in other words, it depends on the axial position of the control rod. Since the sole reason for the return of the plunger is to draw fuel into the space above it, this whole stroke can be referred to as the induction stroke.

For starting some engines from cold, injection needs to be retarded between about 5–10 deg crankshaft angle. This can be done simply by machining a slot in the upper edge of the plunger so that, if the control rod is moved beyond its normal limit set by the maximum fuel stop, the base of the slot, instead of the upper edge of the plunger, closes the inlet port. When the engine is being started, therefore, some sort of mechanical, hydraulic or electrical mechanism is required to disengage the maximum fuel stop. Incidentally, the force that the stop has to react can be fairly large.

Most countries have legislation stating that the excess fuel device must automatically disengage when the engine begins to run normally, so that the maximum fuel stop limits the fuelling. If cold start enrichment is selected manually, the driver must have to leave his seat to bring it into operation

INJECTORS AND IN-LINE AND UNIT INJECTION PUMPS

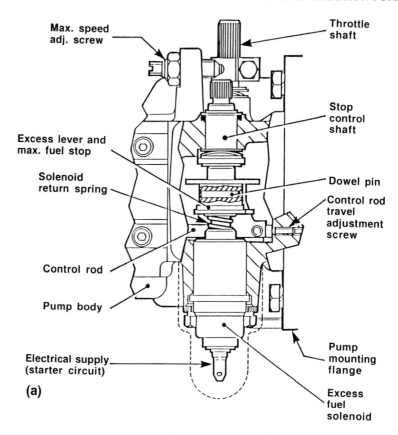

(a)

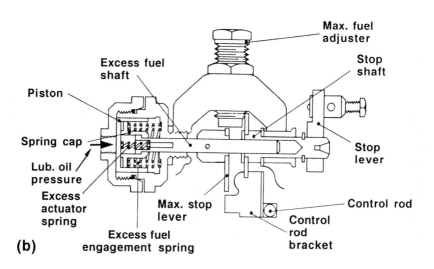

(b)

INJECTORS AND IN-LINE AND UNIT INJECTION PUMPS

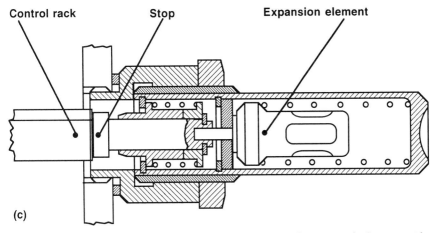

Fig. 5.17 Three methods of resetting the stop so that the fuel rack can move further to provide excess fuel for starting: (a) the Lucas solenoid actuated system, (b) Lucas also make this hydraulically actuated unit in which, when the engine starts, the increasing oil pressure forces the piston to the right to disengage excess fuel (c) the Bosch thermostatically actuated device

and, again, it must disengage automatically. This is to prevent the use of this control for obtaining extra power under normal running conditions, which would cause black smoke to issue from the exhaust. A simple automatic device is used in the Lucas Minimec pump, as explained in the penultimate paragraph of Section 5.18.

Alternatives used by various manufacturers include automatic enrichment by means of a stop the setting of which is determined by a solenoid, Fig. 5.17(a), or a hydraulic, Fig. 5.17(b) or thermostatic device such as a wax-filled element. If a wax-filled expansion element moves the stop, for excess fuel, a spring returns it to its normal setting, as can be seen from Fig. 5.17(c).

5.14 Some plunger arrangements on other pumps

Several variations of the plunger design are practicable. For instance, it is possible to vary the cut-off point by inclining the top edge of the plunger and, lower down, incorporating a peripheral spill groove in a plane normal to the axis of the plunger to obtain a constant spill point. Another variant is a plunger with a combination of both the upper and lower inclined edges. With this arrangement it is possible to vary, as a function of the load, both the spill and cut-off points simultaneously: in other words, both the start and end of injection. This may be desirable, for example, to reduce noise or exhaust emissions.

A different design of plunger, without a counterbore in its end, has been produced by Bosch. Instead of having an inclined straight groove, part of the periphery of the plunger is machined away to form a helical upper edge

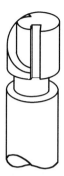

Fig. 5.18 Bosch have produced pump plungers the upper edges of the spill grooves of which are of helical form

and a circumferential lower edge, as shown in Fig. 5.18. A vertical groove interconnects the pressure chamber above the plunger with the peripheral groove above the lower edge. Its function is the same as that of the counterbore in the Lucas plunger, which is to link the spill gash to the pumping chamber. To shut the engine down, the plunger is rotated to the position in which the spill port opens before the inlet port has closed. Bosch also produce plungers with straight gashes for determining the spill and cut-off points.

5.15 Engine control

The timing of cut-off and spill determines the quantity of fuel supplied and therefore the torque and power output at any given speed. When the driver depresses or eases back on his accelerator pedal to increase or decrease the power available to him for accelerating or decelerating his vehicle respectively, or for climbing or descending inclines, he does not necessarily directly move the control rod in the pump. Instead he may do so through the medium of the governor, to which his accelerator pedal is linked.

5.16 Governing and governors

Owing to viscous friction losses and other phenomena outlined in Section 5.17, para 7, the rate of flow of air into the engine does not increase linearly with speed. Consequently, if the rate of fuelling were to be simply increased linearly with speed, or indeed according to any other simple law, it would not match the engine requirements, and the result would be unacceptable levels of undesirable emissions. This is why control by means of a simple linkage between the accelerator pedal and injection pump would be inadequate. Instead, as mentioned in Section 5.15, a governor has to be used to interpret the driver's requirements and to satisfy them within the limits imposed by the engine performance envelope formed by the maximum and minimum speeds obtainable and the shape of the torque curve.

The functions that a governor may be required to perform are as follows:

(1) To supply extra fuel for starting with the engine at any temperature likely to be experienced in service
(2) Maintain the idling speed constant at a predetermined rev/min regardless of temperature and load
(3) Prevent the engine from over-speeding if the accelerator is depressed suddenly at light load or the load is suddenly reduced
(4) An all-speed governor maintains, within pre-set limits for full and zero load, a constant speed appropriate to the accelerator pedal angle, regardless of variations in load
(5) In some instances, a torque control requirement may be added, most frequently to provide torque back-up, by arranging for the torque to increase as the speed falls from the maximum to a pre-set level. However, torque control may also be applied at the other end of the speed range.

Two categories of governor are in general use for automotive applications. The first is the *two-speed governor*, which regulates both maximum speed and that for idling. Between these two conditions, the accelerator pedal is linked directly to the control rod, or rack, so the driver has direct control over the fuelling. Secondly, an *all-speed* governor controls the speed throughout the range, so the driver or operator exercises control only indirectly through the governor, which keeps the speed constant for any given pedal angle. This type of governing is suitable for agricultural tractors, industrial applications and vehicles such as road sweepers and fire fighting equipment, often operated at constant speeds. There are also combinations of the previously described types, obtained simply by using different sets of governor springs, as will be described later.

All-speed governors change fuelling from cut-off to maximum delivery and vice versa over relatively small increments of pedal travel. This can cause not only discomfort to the occupants of the vehicle, but also wheelspin on slippery surfaces, and it is difficult to tolerate with automatic transmission controls. Consequently, two-speed governing is more suitable for most modern automotive applications.

Two-speed governors, by limiting maximum speed, prevent damage that could occur if the engine were to over-speed and, by regulating idling speed, maintain idling stability regardless of engine temperature and load. Even between these two conditions, when the driver is directly in control of the fuelling through the medium of his accelerator pedal, the rate of change of fuelling over the range of movement may be modified by appropriate choice of geometry for the levers and links.

In all governors, some sort of lost motion device generally has to be introduced in the linkage between the governor and the control rod in the pump, to limit the loads applied to the mechanism when the driver calls for an output higher than the limit set by the maximum fuel stop. Within the limits of this lost motion device, the maximum speed governing spring takes over and, provided the accelerator pedal is held in one position, maintains a constant speed regardless of load changes. When the accelerator pedal is released for idling, the control rod actuating mechanism comes back against another spring, which maintains a constant idling speed by appropriately adjusting the fuelling to cater for variations in the load, generally due to switching ancillary equipment on or off, at that end of the speed range. Over the ungoverned middle range, the geometry of the linkage between accelerator pedal and control rod determines the shape of the torque characteristic.

In the context of governing, the term *speed droop* means the accuracy with which the governor controls the speed. It is generally expressed as the percentage increase in engine speed when load falls from the maximum to zero with the accelerator pedal angle constant. For many industrial applications a small droop, typically about 4%, is needed to maintain speeds sensibly constant regardless of load changes. In automotive applications, however, this would lead to jerky performance during load changes so, for good driveability, a larger speed droop, generally about 10–15%, is required.

5.17 Governor mechanisms

Several different types of of governor control have been employed for road vehicles. Pneumatic governing has not been used for many years, because it depends upon manifold depression for exercise of control. It therefore involves throttling the incoming mixture and consequently loss of volumetric efficiency. Furthermore, it tends to be inconsistent in operation and therefore especially liable to be subject to tampering by drivers. In any case, if the engine were to stall, the governor could not prevent it from running uncontrollably backwards.

Three types of governing are now available: they are mechanical (centrifugal), electronic only and electronic with hydraulic assistance. Examples of these systems are illustrated and described in Sections 5.18, 5.19 and 5.20. To satisfy emissions regulations the trend seems inevitably to be towards electronic control.

Mechanical governing systems generally comprise a set of weights (numbering between 2 and 6) radially disposed about, and rotating with, the input shaft to the injection pump. They are mostly carried on the ends of bell-crank levers the other ends of which bear against one end of a sleeve, to slide it axially along the shaft. The other end of this sleeve generally bears against, or is attached to, the lower end of a vertical lever that actuates the fuelling control rod, as illustrated in Figs. 5.13, 5.20 and 5.24.

Centrifugal force, increasing as the square of speed, tends to throw the weights outwards, and therefore to move the sleeve to one extreme of its travel. This movement is resisted by a spring, or a combination of springs, tending to move it to the other extreme. Consequently, the setting of the pump control rod is determined by the position of equilibrium of the mechanism under these two forces.

Various spring types and layout are employed, some of which are described later. A single coil spring in a pocket in each weight, and constrained by a bolt, Fig. 5.19(a), has the disadvantage that the linkage connecting it to the control rod tends to become complicated, owing to the need to tailor the motion to match the fuelling requirements of the engine. If slider blocks are incorporated in the linkage, as in this illustration, friction can be a problem. Another arrangement widely used is the introduction of coil springs in the linkage between the weight mechanism and the control rod, Fig. 5.19(b). Alternatively a coil spring or springs may bear against one end of the sleeve. Leaf springs have been used as in Fig. 5.20(a), and torsion springs as in Fig. 5.20(b). Two or more springs of any of these types may be arranged to come into effect sequentially and thus to vary the axial motion characteristic of the sleeve. Other examples and illustrations will be found in Chapter 6.

With the bell-crank lever arms cast integrally with the pivoted weights, the mechanism is extremely simple and has a small number of parts. Given appropriate geometry, the lever arms can become effectively part of the mass and thus, by contributing to the total centrifugal force, reduce the overall diameter of the mechanism. Also, the geometrical layout can be designed to modify the control characteristics of the governor. A further reduction in diameter can be obtained by interposing a pair of gears between the camshaft and the governor, to increase its speed of rotation and thus enable smaller weights to be used. However, this tends to be both heavy and costly and therefore tends to be confined to governors for slow speed engines to increase the output forces from the weights.

Although governors were originally introduced for controlling the speed, the practice of adding a torque control function to optimise the performance of the engine, mainly at the upper end of its speed range, is becoming increasingly widespread. Torque control modifies the full-load fuel delivery. It is desirable for naturally aspirated engines because, as engine speed increases, the rate of increase in air inducted tends to fall off relative to the rate of fuel injected. This is partly because of viscous drag and the throttling effects of obstructions such as poppet valves, guides etc, on the air. Other factors apply too, such as thermal effects and variations in the quality of mixing of the fuel and air with rate of fuel flow through the injectors.

In Fig. 5.21, curves of engine fuel delivery requirement plotted against speed are compared with actual fuel delivery with and without torque control. To match the delivery characteristics, it is necessary to reduce progressively the rate of advance of the rack in an in-line pump.

With a turbocharged engine, the full load fuelling requirement over part of the lower speed range generally increases more rapidly than the normal output from the pump. In this case, it will be necessary to increase fuelling up to a certain speed, and then reduce it again, Fig. 5.21. These effects can be obtained either by inserting an extra spring to interact with the main governor spring, or by the use of a boost pressure actuated diaphragm mechanism Figs. 5.22 and 5.26, which moves the maximum fuel stop to increase the travel of the control rod.

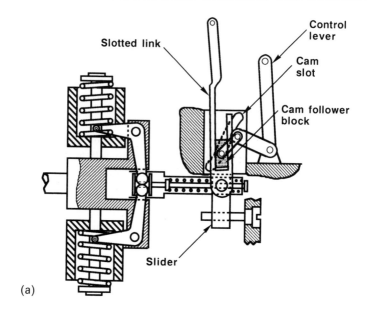

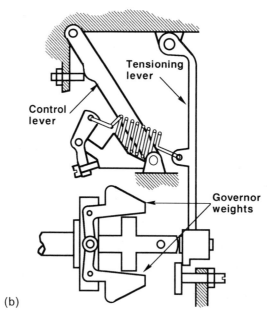

Fig. 5.19 At (a) is shown diagrammatically a Bosch governor mechanism design, with coil type governor springs housed in the weights, and a cam follower which slides in a slot in a lever and is guided by a cam slot. At (b) is an alternative Bosch design in which a coil type governor spring acts, through the medium of a tensioning lever, upon the governor weight mechanism

INJECTORS AND IN-LINE AND UNIT INJECTION PUMPS

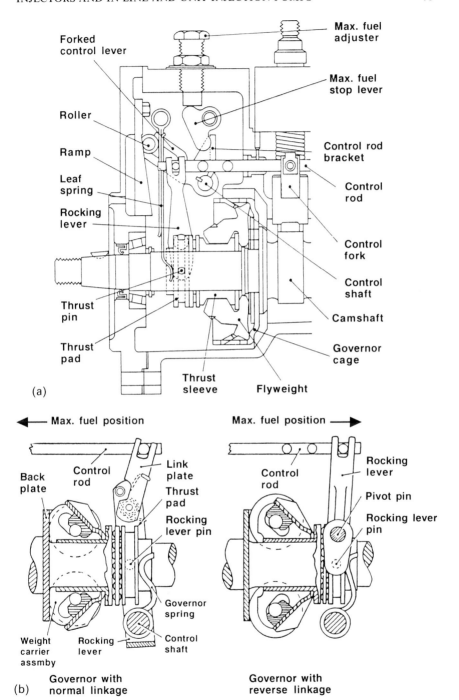

Fig. 5.20 The Lucas Minimec governor (a) has a leaf spring. At (b) is shown a Minimec governor with a coil spring loaded in torsion

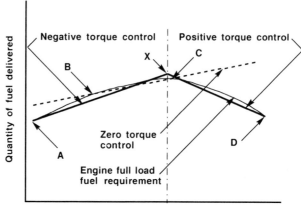

Fig. 5.21 The curve ABCD shows the full load fuel requirement for a turbocharged engine and the dotted line the fuel supply as it would be without governing. Superimposed on these are negative and positive torque control characteristics, meeting at X. It is possible to get closer to the ideal by applying negative torque control between A and B, zero torque control between B and C, and then positive torque control between C and D

5.18 Mechanical governors produced by Lucas

Fitted to the Minimec pump is the C type governor, illustrated in Figs. 5.13 and 5.20. Two, three, four or six weights can be carried in the cage. For accommodating either three or six weights, there are six pockets, Fig. 5.23. In all installations, the weights pivot in the corner between the the flange and back of the cage. The cage is bolted directly to a flange on the camshaft. As an option, a cush drive can be interposed between the two to absorb torsional vibrations of the shaft.

To convert the radial force on the weights into an axial force, fingers projecting from the weights abut against a thrust washer interposed between them and a sleeve sliding axially on the shaft. This force is transmitted through a needle roller thrust bearing, to a collar, termed the thrust pad, which therefore slides axially with the sleeve. As can be seen from the illustration, the lower end of a quarter-elliptic compound leaf spring bears against the thrust pad, thus tending to oppose the motion induced by the centrifugal force on the weights. This spring is pivoted at its upper end, and a movable stop is interposed between it and a ramp on the housing wall.

The movable stop, the function of which is to modify the spring force, is in the form of a pair of rollers interposed between it and a ramp, or face-cam, mounted on the wall of the housing. These rollers are carried on the ends of a pin, which is itself carried between the arms of a fork attached to the spindle that is rotated by the speed control lever actuated by the accelerator pedal.

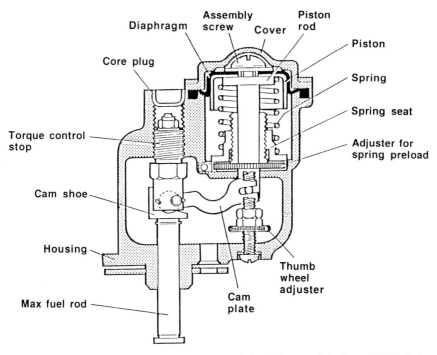

Fig. 5.22 Increasing boost pressure acting on top of the diaphragm of the Lucas BC200 Series boost control unit forces the piston progressively downwards the until the transverse pin in the lower end of its stem abuts against the end of the thumb wheel adjuster. This motion causes the cam plate to pivot about the small diameter pin near its opposite end and the adjacent eccentric cam therefore to lift the stop for the fuel rod, to increase the rate of fuelling

Rotation of the fork, by increasing or decreasing its effective length, moves the pin and roller assembly up and down the spring, and therefore varies its rate. Additionally, as it moves up and down the ramp it, in effect, varies also the preload on the spring, changing the governed speed of the engine. Simply by fitting different ramps, the governor characteristics can be made to suit the requirements of different engines, as regards idling and limitation of maximum speed. Registering in a groove around the thrust pad, and therefore moving with it, is a pin projecting from the lower end of a centrally pivoted vertical lever, termed the *governor lever*. This lever signals to the control rod, or rack, the position of the thrust pad.

Tamper-proof provision of excess fuel is obtained by the use of a latch mechanism, which can be manually engaged by the driver only if he leaves his seat. The latch temporarily holds the control rod in the excess fuel position and, as soon as the engine speed rises after starting, automatically springs out of engagement.

With many governors, the torque curve can be made to fall as speed increases by simply replacing the usual maximum fuel stop with a spring.

 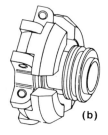 

(a)   (b)   (c)

Fig. 5.23 Three alternative weight units for Minimec and Minimec 9 mm governors: (a) GE low cost unit for tractors and other uncritical applications (b) GX precision unit for critical governing applications other than the 9 mm pump (c) available with rigid and cush drives and designed to combine the good features of the GE and GX weight units. The specifications are as follows, the maximum to no load run up figures being quoted for BS649:1958, Class 2 conditions:

	Weights	Pump speed range, idling to max.	Min. full load to no load run up	Max. control lever force
(a)	2	250–1500 rev/min	8%	4.9 NM
(b)	2 or 4	250–2000 rev/min	4.5%	6.9 NM
(c)	2, 3 or 6	225–2500 rev/min	4.5%	6.9 NM

This provides a torque back-up characteristic that is commonly required for engines to be installed on vehicles such as diggers and some tractors, which must not stall when a heavy load is suddenly applied.

5.19 Bosch in-line pumps

Among the in-line pumps produced by Robert Bosch GmbH is the P Series, designed for direct injection engines, producing up to about 70 kW/cylinder, for installation in heavy duty truck engines.

Longitudinal and transverse sections of the Bosch P-7000 series in-line pumps are illustrated in Figs. 5.24 and 5.25. Its mechanical governor, which operates on the principles described in Section 5.17, is shown in the longitudinal section. The governor springs are housed in counterbores in the outer ends of cylindrical weights, the radial motions of which are converted into axial motion and transmitted to the end of an angle-section control rod by a complex system of levers and links illustrated diagrammatically in Fig. 5.19(a). Incidentally, Bosch also produce governors with integral weights and levers and having separate springs.

With the mechanical governor, excess fuel for cold starting can be obtained automatically by means of a variable stop limiting the movement of the control rod towards its normal maximum fuel stop. The variable stop is moved by the expansion and contraction of the wax in a thermo-actuator, in response to changes in temperature.

INJECTORS AND IN-LINE AND UNIT INJECTION PUMPS

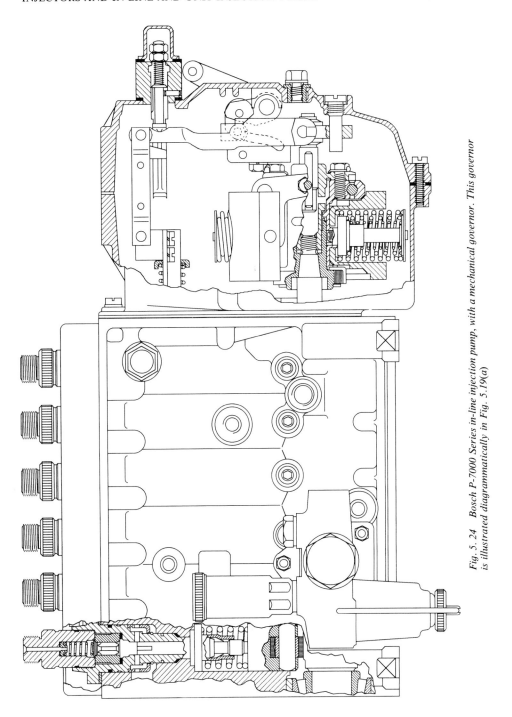

Fig. 5.24 Bosch P-7000 Series in-line injection pump, with a mechanical governor. This governor is illustrated diagrammatically in Fig. 5.19(a)

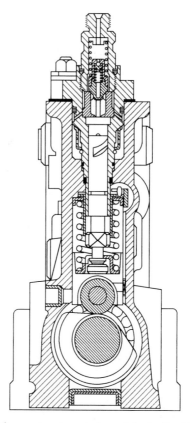

Fig. 5.25 Cross section of a pumping element of one of the the Bosch 7000 Series in-line injection pumps

Mounted on the opposite end of the pump is a manifold pressure compensator, termed by Bosch the *LDA*, for turbocharged engines, Fig. 5.26. A diaphragm subjected to manifold pressure actuates a push rod, that bears on one end of a bell-crank lever. Carried on the other end of the bell-crank lever is an adjustable screw stop that limits the motion of the control rod towards its maximum fuelling position. This device, by moving the maximum fuel stop, automatically increases the fuelling as the boost pressure rises. A similar mechanism can be used to compensate for altitude.

5.20 Bosch electronic control

The EDC (electronic diesel controller) governor produced by Bosch to replace mechanical governors is illustrated in Fig. 5.27. Its actuator is a solenoid which exerts, against the resistance of a return spring, a maximum force of

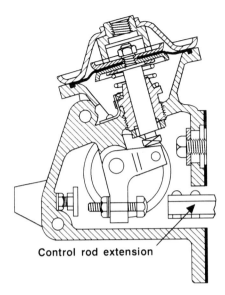

Fig. 5.26 The Bosch LDA manifold pressure compensator for turbocharged engines

45 N directly on the end of the control rod. The current through the solenoid is regulated, in response to feed-back signals received by the ECU.

The electronic control system (ECU) is illustrated in Fig. 5.28. Sensors transmit signals to the ECU to provide it with some, though rarely all, of the following data: accelerator pedal position; engine and road speeds; gear selected; start of injection; control rod position; and ambient, induction manifold, and fuel temperatures; ambient and boost pressures; crankshaft angle; and clutch, brake pedal, and exhaust brake and retarder switches. The outputs from the ECU are generally the current to the solenoid for actuating the pump control rod, and that to the injection advance and retard mechanism.

Signals of accelerator position indicate to the microprocessor in the ECU what torque is demanded by the driver, and the control rack position indicator signals the current torque setting. Engine speed signals are supplied by an electronic pulse generator mounted, below the solenoid, on the end of the pump camshaft. On the basis of all these pieces of information, plus the others such as temperatures and pressures, the ECU compares the position of rack with a map showing what its position it should be to provide the required torque. The ECU then modifies the current passing through the solenoid, to alter the position of the rack appropriately.

This system has a number of advantages which include: engine starting and stopping done solely by means of a key on the dash fascia, regardless of engine and ambient temperatures; idle speed adjustment to cater for changes in load from auxiliary equipment; relative ease of adjustment of full-load fuelling to keep within smoke limits; automatic adjustment of fuelling

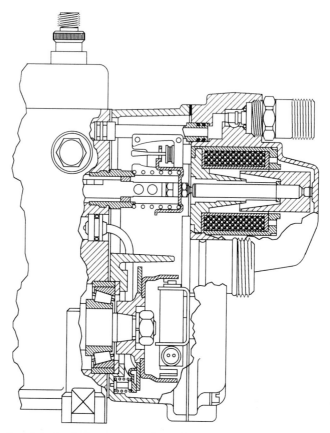

Fig. 5.27 One of the Bosch electronic control units for in-line diesel injection pumps. The position of the stop for the rack, or control rod, is varied by an electronically controlled solenoid acting against the return spring. A rack position sensor is mounted immediately above the end of the rack and the return spring assembly and the pump speed and timing sensor is below, on the end of the pump camshaft

in relation to ambient, fuel and coolant temperatures; ease of controlling auxiliary drive speeds; ability to incorporate surge damping and cruise control and to generate signals for indicating, on the dash fascia, fuel consumption and engine speed. It is also possible to program self-diagnosis into the ECU.

Bosch, working in association with Daimler-Benz AG, have also developed an electro-hydraulic actuator, Fig. 5.29, for the D-B OM 422 LA engine. In principle, the control mechanism is similar to that of the electric actuator but a much smaller solenoid housed in the end cover displaces a spool valve axially, against the force exerted by its return spring. The movement of this valve directs fuel under transfer pump pressure either into a chamber between a piston and the end of its cylinder or to release it from the chamber back to the fuel tank. The piston, capable of exerting a force of 90 Newtons, is

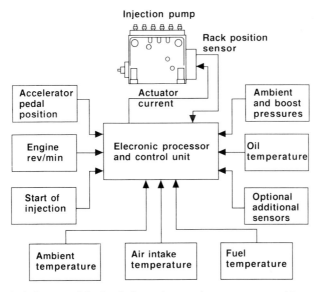

Fig. 5.28 Block diagram of the Bosch electronic control system system and its sensors

mounted on the end of the control rod, which it moves against the force exerted by its return spring.

5.21 Unit injection

Unit injection implies the use of a combined injection pump and nozzle for each cylinder. It has the advantage that the long, high pressure pipelines from pump to injector are absent, and so also therefore are the compressibility and wave effects associated with them. Consequently, very high pressures are practicable. As explained in Chapter 7, smoke can be reduced by increasing the injection pressure which, by reducing the degree of air swirl needed, improves volumetric efficiency and reduces heat loss to coolant. Therefore, unit injection also offers the additional bonus of a potential reduction in fuel consumption.

Among the other advantages of this type of injection equipment is that a single unit can be produced for application to a wide range of engines having different numbers of cylinders and power outputs. Servicing is simplified because removal and replacement does not entail retiming. Because the high pressure ducts are not only short but also of equal length, and are drilled in the body of the injector, the rate of fuelling is consistent from cylinder-to-cylinder, the risk of leakage is virtually nil, any wave effects that do occur are at very high frequencies and low amplitudes, and timing can be more accurate and consistent.

The pump can be actuated by cams on the shaft serving the overhead

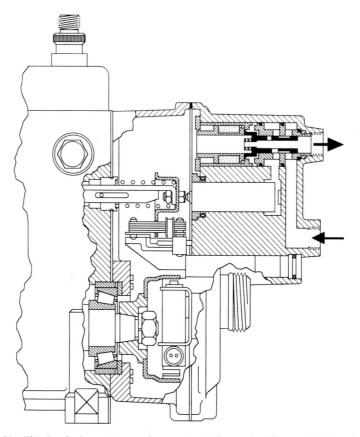

Fig. 5.29 This Bosch electronic control system is similar to that illustrated in Figs. 5.27 and 5.28, but has a hydraulic servo. Fluid under pressure entering by the lower connection is directed by the solenoid-actuated valve, top right, either down to the chamber behind the piston immediately below, or released through the upper connection. Mounted on the other end of the piston is the stop for the rack. The rack position sensor is below the rack return spring assembly

valve gear, though the diameter of the shaft must be increased, to cater for the high loading applied to it by the injectors, Fig. 5.30. Alternatively, it can be actuated by pushrods from a camshaft carried high on one side in the cylinder block casting, Fig. 5.31. Short pushrods are essential, otherwise the mechanism will not be stiff enough. In the absence of adequate stiffness, the injection characteristics obtainable may not be much better than those of the more traditional types of pump. Unit injection is generally at its best in a four-valve head, Fig. 5.32, because the axis of the nozzle can be coincident with that of the cylinder, and uniform spray distribution is therefore relatively easy to obtain. Because of the heavy loading, roller type cam-followers are almost invariably used.

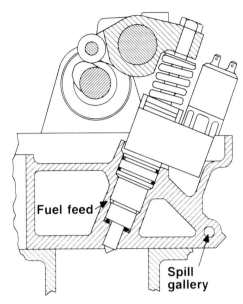

Fig. 5.30 A typical camshaft-actuated unit injector layout for an engine with two valves per cylinder

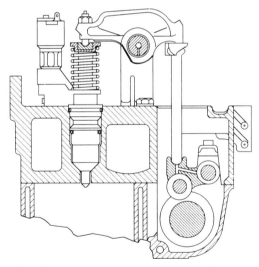

Fig. 5.31 With push-rod and rocker actuation, the unit injector can be accommodated vertically and inject symmetrically into the centre of the combustion chamber

On large medium and slow speed diesel engines, unit injection has been employed for many years. However, external piping has been widely used to take the fuel at low pressure to the injectors. Moreover, the governor has exercised control through the medium of linkage extending almost the whole

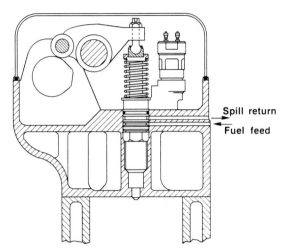

Fig. 5.32 A simple layout for a cam-and-rocker actuated unit injector. Note that, because of the very heavy loading on the cams, roller followers are the norm, as indicated in this and the two previous illustrations

length of the cylinder head, and large forces are needed to actuate these linkages. The injectors are bulky and accommodating both them and the mechanical control is often difficult.

On high speed engines, advancing the injection timing with increasing speed is essential for the maintenance of combustion efficiency, and this is difficult with mechanically controlled unit injection. Mechanisms similar in principle to those described in connection with variable valve timing in Chapter 10, might be used but are bulky, costly and complex. Electronic control therefore has to be considered.

5.22 Electronic unit injection — Lucas EUI System

Increasing stringency of emissions regulations have led manufacturers to the conclusion that precise timing control with very high rates of injection with a sharp cut-off are needed. Such high rates of injection entail pressures of up to 1500 bar, which are hardly practicable with either in-line or distributor type injection pumps. This is principally because large spill port areas are needed to obtain an adequately rapid collapse in pressure, and this would lead to cavitation erosion in the relatively long high pressure lines to the injectors.

For all these reasons, Lucas decided in the early 1980s to develop their EUI (Electronic Unit Injector) system, which has been the subject of several papers. Among these are 'Electronic Unit Injectors' and 'Electronic Unit injectors Revised', SAE Papers 885013 and 891001, both by Frankl *et al.* For such high performance injection equipment, electronic control was clearly

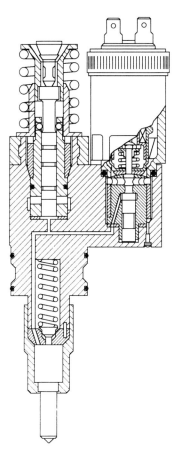

Fig. 5.33 This Lucas electronically controlled unit injector has a cam-actuated pump, and control is effected by a solenoid-actuated spill valve

desirable, and had the additional advantage that it opened up the possibility of interaction with other drive-train control systems.

Since unit injectors are accommodated inside the valve cover, leakage of fuel could dilute the lubricant, so it has to be avoided. Consequently, for the Lucas system, both the feed and spill galleries are incorporated in the cylinder head casting, as can be seen in Figs. 5.30 and 5.32.

In the body of the Lucas injector, Fig. 5.33, three units are accommodated. The nozzle is assembled up to it from below; the cam-actuated pumping unit is housed immediately above it; and an integral extension overhung to one side houses a solenoid-actuated spill valve. Actually a Colenoid (a patented form of solenoid having a stepped conical core) is employed. Two bolts, passed through holes in a flange at the upper end of the part of the housing that contains pump element, secure the whole assembly to the cylinder head.

These bolts pull a conical shoulder around the injector body down on to a conical seating which forms the gas seal in the head. Liquid sealing (both oil and fuel) is assured by two O-rings in grooves around the base of the housing that contains the injector spring.

The principle of operation is clear from Fig. 5.33. Except when pushed down by the cam, the pumping plunger is fully retracted by the spring above it. At the same time, the spill valve is held open by its own spring. Fuel is supplied, by an engine-driven, low pressure pump, to the feed gallery. It then flows continuously on through the feed port into the pump chamber, and thence through the spill valve back to the tank. This helps initially to prime the system and then to clear it of any air that might have entered.

As the plunger is pushed down by rotation of cam, its lower edge closes the feed port. Consequently, the further downward movement continues to displace fuel through the spill port until the solenoid is energised to close it. At this point, the pressure in the pumping chamber rises rapidly until it lifts the injector needle off its seat, and injection begins. Termination of injection is signalled by the collapse of the electric field in the solenoid, allowing the spill valve to be opened by its spring.

The electronic control system is illustrated in Fig. 5.34. Accelerator pedal position and angle sensors signal to the ECU the demands from the driver. An inductive transducer picks up signals from a 60-tooth wheel mounted on the crankshaft or flywheel, which are used as the basis for calculating speed, timing and fuelling. This sensor also provides the indications of crankshaft angle, on the basis of which the solenoid is switched on and off to trigger the start of injection and to determine its duration. Another wheel, but driven by the camshaft and having one more tooth than the engine has cylinders, indicates which injector is to be activated, the extra tooth marking the end of one and the beginning of the next series of cylinder fuelling operations.

Signals of engine temperature initiate cold start routines and, additionally, are compared with a map for temperature compensation of fuelling. Turbocharged engines are equipped with a boost pressure sensor, to increase fuelling progressively over the boosted range of operation, but to limit it during acceleration to compensate for the fact that, during this mode of operation, the air delivery tends to lag behind fuel delivery.

The tolerances on accuracy of timing and quantity of fuel supplied are very tight, a shot-to-shot consistency of about 5 μsec having to be maintained. Either a 12 or 24 volt battery supply can be utilised, and internal power units provide 90 volt feeds to the Colenoid and microprocessor circuits. Among the advantages of such a high voltage, relative to that of the battery, are the following. The currents involved are low; smaller gauge wiring can be used in the harness; performance is less sensitive to variations in resistances of wiring and battery voltage; and, as an overall consequence, fuelling consistency is better.

Electronic control offers the benefits of extreme flexibility as regards timing, torque shaping, and governing strategies: for example, it is easy to provide either two- or all-speed governing and special features such differing speed-droops for different applications. Moreover, with this system, the

INJECTORS AND IN-LINE AND UNIT INJECTION PUMPS

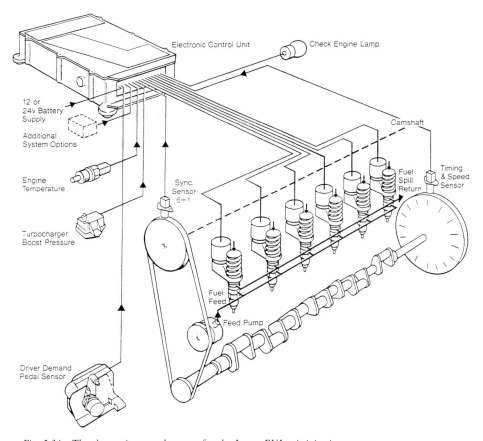

Fig. 5.34 The electronic control system for the Lucas EUI unit injection system

parameters for each injection are entered and calculated immediately prior to the event. Consequently, the responses to changes in demand are extremely rapid. Indeed, it is not impossible to change from zero to maximum fuelling between two injections. Also it is possible to embody adaptive control of such accuracy that compensation can be made for cyclic variations. The outcome is smoothly progressive control and good driveability.

Tests have shown that the variations in the interval between the start of the logic pulse and the start of injection is only 5 μsec, which is the equivalent of about 0.1 deg crank angle. Consequently, any inaccuracy in timing is likely to be due not to the electronics but to incorrect setting of the toothed wheel and transducer relative to tdc of the piston in No. 1 cylinder. Over a period of 10 μsec, 1 mm^3 of fuel is injected, and a tolerance of $\pm 4\%$ is specified for all units. Most of this tolerance is to cater for variations in flow past the nozzle tip.

Temperature is another factor that can affect fuel delivery. This is because, at the high pressures involved, the bulk modulus of the fuel varies with temperature, Fig. 5.10(b). For this reason, it is important to ensure that temperatures of fuel delivered through the low pressure ducting in the cylinder head and delivered to each injector do not differ significantly.

5.23 The GM unit injector

Among the early unit injection systems is that which was introduced by GM Rochester. In 1988, however, this injection business was sold to a new company set up for continuing its development and production. Originally the new company was jointly owned by Penske Transportation Corporation Inc, whose holding was 80%, and Detroit Diesel Corporation, with 20%. Currently, however, the ownership is divided Penske Transportation 51% and Robert Bosch GmbH 49%. Production has continued in the former Rochester Products plant at Wyoming.

As in conventional in-line type injection equipment, pumping and fuelling control are effected respectively by the individual plungers and their spill grooves, Fig. 5.35. The plungers are rotated by a governor and control rod assembly, the latter comprising a rack comprising as many interlinked segments as there are cylinders on the engine. Teeth machined on each segment of the rack mesh with those of pinions machined on the plungers, but not shown in Fig. 5.35.

The fuel is lifted from the tank by a gear type feed pump, through the usual supply system comprising an optional water separator and coarse and fine filters, to a gallery pipe extending most of the length of the cylinder head. From the gallery, branch pipes take it to the injector units. On entering the injector body, the fuel is passed through a fine filter into a transversely drilled hole. From the inner end of this hole, it passes vertically down through another, and then on through a vertical slot in the periphery of the barrel, into a space between the barrel and a sleeve, termed the *spill deflector*, surrounding it. This annular space is termed the *fuel supply chamber*. Radial holes one above the other in the plunger barrel, form the inlet and spill ports.

Fuel in excess of requirements passes upwards, through the inclined hole in the barrel, from the sleeve into an annular groove in the bore of the barrel, whence it passes through vertically and transversely drilled holes, not shown in the illustration but similar to those for the inlet, to the outlet connection and back to the tank. This flow serves for cooling the injector and keeping it free from air bubbles. The outlet is smaller than the inlet, so that the fuel in the body of the injector is always under pressure.

As the plunger is lifted by its return-spring, after an injection, it opens the inlet port. At this point, fuel at feed pump delivery pressure passes in through the inlet port, into the spill groove, termed the *fuel metering recess*, and on through radial and axial holes into the base of the plunger to fill the space beneath it. Incidentally, the spill port too may be open during part of the inlet stroke of the plunger, dependent on the torque setting of the rack: the higher the torque setting, the shorter is the time that it is open. When it is

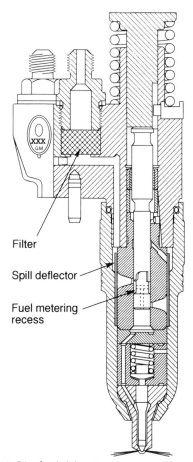

Fig. 5.35 The GM Detroit Diesel unit injector

open, fuel can flow in through it and also pass down the axial hole in the plunger to help to fill the space below. At one setting of the rack, which is that for stopping the engine, the spill port is open throughout the stroke of the plunger.

As the plunger is forced down again by the rocker type cam follower, some of the fuel beneath it is displaced through the lower port into the annular fuel supply chamber around the barrel, until the lower end of the plunger closes this port. Subsequently, the fuel continues to move up from beneath the plunger, through the axial and radial holes in it into the fuel metering recess, and out again through the upper port into the fuel supply chamber until the upper port is closed by the helical edge of the fuel metering recess. At this point, the fuel remaining beneath the plunger is subjected to rapidly increasing pressure. As the plunger continues its downward movement, first the check valve opens, allowing the fuel to pass though it and build up the pressure in the spring chamber and tip cavity until it is high enough to lift the needle valve off of its seat. At this point, injection begins. Both the opening and closing of the needle valve are extremely rapid because, when it is open, its cross sectional area exposed to the fuel pressure is much greater than that when it is closed. Injection is terminated when the lower land of the plunger uncovers the lower port, thus spilling the remainder of the fuel beneath the plunger back into the fuel supply chamber. A radial hole in the spring housing serves as a pressure relief duct, allowing fuel that has passed at injection pressure between the needle valve and its guide to bleed away, thus obviating any risk of dribble.

5.24 Penske/Detroit Diesel electronically actuated unit injection

In 1984, GM introduced an electronically controlled version of its unit injector. While this is much better suited for complying with the emission control regulations than is the mechanically actuated version, the fact that it has to accommodate a solenoid, renders it less compact. A second generation version of the Detroit Diesel Electronic Control (DDEC II) went into production, by Penske, in late 1988.

Originally, in the (DDEC I), Fig. 5.37, the ECM was used in conjunction with an electric distributor unit (EDU) the function of which was to serve as a high current switching unit to forward the commands to the solenoid. In both, it monitors some of the sensors that send signals to the engine's electronic control module and, by sending command pulses ultimately to the solenoids in the electronic unit injectors (EUIs), controls the timing and quantity of fuel injected into each cylinder. If it receives signals indicating, for example low oil pressure or coolant level, or high oil temperature it limits the output of the engine or shuts it down completely. In the DDEC II system, a fuel temperature sensor is added and the EDU has been eliminated. All functions are therefore centered upon the ECM, which communicates directly with the EUIs.

The hardware is illustrated in Fig 5.36. From this, it can be seen that the fuel enters through two diametrically opposed inlet ports, each containing a filter screen to prevent coarse foreign matter from passing through into what is termed the nut cavity, and on to the injector mechanism. The nut cavity is an annular space around a cylindrical spacer between the injector body and the check valve cage. From this cavity, the fuel passes through a series of holes to the solenoid actuated poppet type control valve through which, when open, it is delivered to the plunger chamber. The control valve determines the plunger stroke by initiating and terminating the fuel supply to the plunger chamber.

Incidentally, the disc type check valve has no part to play in the injection sequence. It is there simply as a safety precaution, to prevent combustion gases from blowing back into the injector and fuel system in the event of the needle valve failing to shut completely owing to debris becoming lodged between its seating faces.

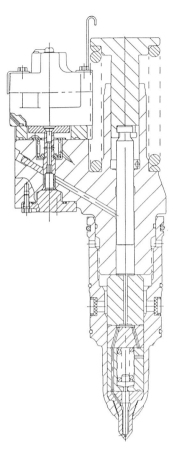

Fig. 5.36 The Penske Corporation have been producing this electronically controlled unit injector with a solenoid-actuated spill valve

An interesting detail is the use of what is termed an *injector tube*, Fig. 5.38, to bring the coolant closer to the tip of the injector than it would be if the cast housing were to be continued right down to the lower deck of the cylinder head. This tube is in fact a copper cup pushed down from the upper end of the injector housing bore until a lip around its upper end closes down on an O-ring seal on shoulder in the bore, and its lower end seals on a spherical seating machined in the lower deck of the casting.

Fuel delivery and return galleries in the cylinder head are connected respectively to the feed pump and fuel tank, and a solenoid-actuated valve performs the spill function. Consequently, there is no spill groove on the plunger. As the plunger rises, fuel (at feed pump delivery pressure) fills the chamber beneath it and, so long as the solenoid valve is open, continues on through a passage up to the solenoid valve, whence, for cooling the unit and ensuring that air cannot be trapped in the system, it is directed through another drilled passage to the fuel return gallery in the cylinder head.

The plunger, actuated by a cam and rocker, begins its downward motion

112 INJECTORS AND IN-LINE AND UNIT INJECTION PUMPS

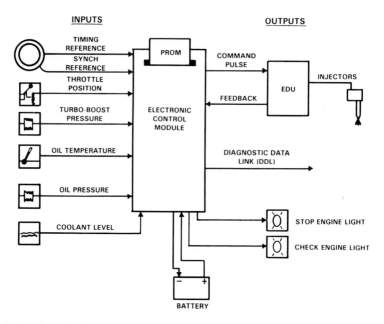

Fig. 5.37 This is the first generation DDEC electronic control system for the GM unit injectors. It differs from the second generation system in that the command pulse and feedback are directed to and from the injectors through an EDU instead of directly. The EDU (electronic distributor unit) functions as a high current switching unit for energising the solenoids

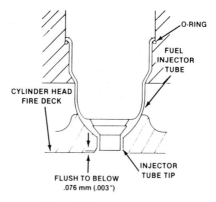

Fig. 5.38 By virtue of the use of this thimble shape tube, the coolant can be brought close to the injector tip

as the engine piston moves beyond about two-thirds of its upward stroke. However, injection does not begin until the ECM has energised the solenoid to close the spill valve, and the pressure beneath the plunger has, as a result, risen high enough to lift the needle valve from its seat in the hole type nose. The injection pulse width is determined by the ECM, which switches off the

INJECTORS AND IN-LINE AND UNIT INJECTION PUMPS   113

current to the solenoid, to allow the spill valve to open under the influence of its return spring. What is known as the response time feedback (the delay between the despatch of the signal to close the spill valve and its actual closure) is monitored by the ECM, so that it can compensate for injector-to-injector variations in timing.

5.25 Cummins PT unit injection system

This system was introduced in 1924. The initials PT stand for pressure-time, the flow of fuel through the fixed metering orifice depending on the fuel pressure and the length of time that it is open. In the diagrammatic representation of the system, Fig. 5.39, items 3 to 9 inclusive are all housed in or mounted on a combined pump, governor and control housing, normally flange-mounted on the timing cover of the engine. In Fig. 5.40, the arrangement is shown more clearly.

The shut-down valve can be either manually or electrically actuated. Basically, it is a spring-loaded spool-valve which, when in the shut-down position, directs the fuel delivery from the throttle valve directly back to the fuel tank, instead of to the fuel manifold.

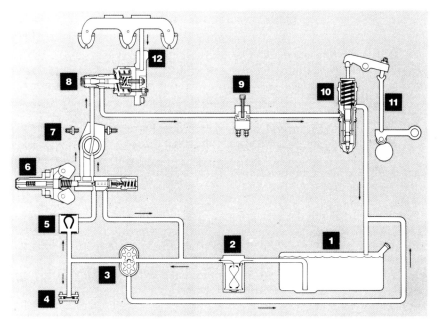

Fig. 5.39 The Cummins PT unit injection system
1 Fuel tank, 2 Filter, 3 gear type fuel pump, 4 pulsation damper, 5 Magnetic screen, 6 Governor regulating pressure, 7 Hydraulic throttle, 8 Air/fuel control valve (AFC) for turbocharged engines, 9 Shut-down valve, 10 unit injector, 11 Injector actuation mechanism, 12 Connection to induction manifold

Fig. 5.40   The Cummins PT governor and control assembly

A shaft from the engine timing and ancillary gears is coupled to the governor drive shaft, into the opposite end of which is splined the shaft driving the gear type feed pump. Both rotate at half engine speed, but the governor drive gear meshes with a pinion on the governor shaft, to drive the latter at engine speed.

The governor is of the conventional centrifugal type, the axial motion of its sleeve moving a spool-valve against the resistance of a coil spring bearing against its opposite end. If two-speed governing is required, a second governor and spool-valve assembly is installed above the drive shaft, its pinion meshing with the same drive gear, Fig. 5.41. In fact, a variety of governors is available for meeting different requirements, including zero speed-droop.

The fuel feed pump delivery pressure is fairly high, about 17.2 bar, which is why a gear type pump is employed. The actual pressure in the manifold serving the injectors is regulated by a hydraulic throttle, which is actuated by a linkage connected directly to the accelerator pedal. Only at the upper end of the speed range does the governor take over, primarily to prevent over-speeding if the load is suddenly released.

Fuel from the tank passes through a filter to the feed pump, which delivers it to a T-junction in the ducting in the feed pump housing. One arm of the T is connected to a pulsation damper, while the other delivers the fuel through a magnetic screen to the governor-actuated spool valve. From this valve, the fuel flows on through a hydraulic throttle and shut-down valve to the gallery serving the unit injectors.

INJECTORS AND IN-LINE AND UNIT INJECTION PUMPS

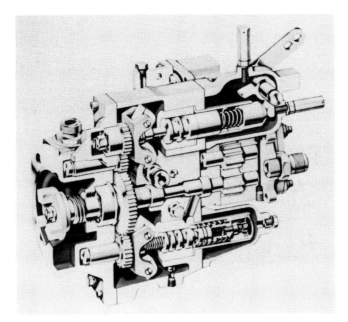

Fig. 5.41 If two-speed governing is required, a second governor and spool valve assembly is installed in the top of the casing of the Cummins unit

To cater for turbocharger lag, what Cummins term an *air/fuel control valve*, or *AFC*, can be interposed between the hydraulic throttle and shut-down valve. The AFC is a diaphragm-actuated, damped spool-valve that limits the rate of increase of fuelling during acceleration, to offset the tendency to emission of black smoke, until the rate of increase of air supply catches up.

5.26 Cummins PT combined pump and injector units

Each injector, Fig. 5.42, comprises two main components termed the *cup* and *cup retainer*. The cup retainer, which might be likened to the pump barrel, contains the plunger. Its lower end seats on the upper end of the nozzle which, in turn, seats on a tapered shoulder in the base of the cup. At its upper end, two bolts, passed through holes in a saddle bracket, secure it to the cylinder head. There is a clearance between the vertical walls of the cup retainer and cup, so that the plunger, cup retainer and cup can centre freely relative to each other. At the upper end of the plunger is its return spring. As the plunger is moved downwards towards, by the cam-and-pushrod actuated rocker, this spring is compressed, subsequently lifting the plunger again as the follower passes the nose of the cam.

In Fig. 5.43, the injector is shown diagrammatically to a larger scale, to illustrate the stages of its operation. Before the plunger begins its upstroke,

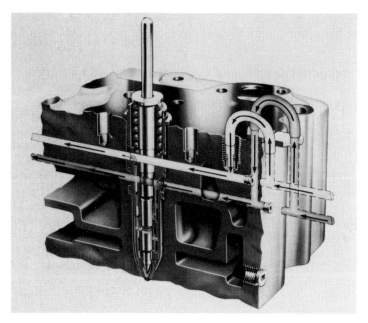

Fig. 5.42 Section of the Cummins PT unit injector installation in a cylinder head

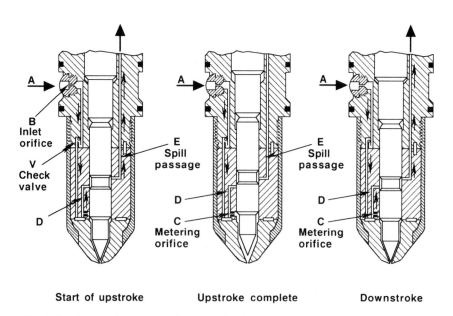

Fig. 5.43 Showing the sequence of events in the Cummins unit injector

fuel from the manifold passes through the inlet port A and the flow control orifice B. From this orifice, the fuel goes down through a drilled duct, turns up to pass through a check-valve, and then continues down again through passage D into an annular groove in the lower end of the cup retainer. This groove is closed by the flat upper end of the nozzle, on which the cup retainer seats.

From groove, the fuel flows upwards again, into a vertical duct into which two radial holes are drilled. Since the lower of these two holes, which houses a metering orifice, is closed by the plunger, the upper one delivers the fuel into an annular groove around the plunger, through which it flows up through duct E and back to the tank. This flow, the rate of which is determined by the the fuel pressure and size of inlet orifice B, is for cooling the unit. The fuel pressure is a function of the engine speed, governor and throttle position.

As the the plunger moves upwards, the lower radial hole is uncovered, and fuel flows, through the metering orifice into the injector cup. The quantity of fuel entering the cup is a function of the size of the orifice and the pressure. At the same time, the upper of the two radial holes is covered by the plunger, so the circulation of fuel back to the tank ceases. Any tendency for blow-back to occur momentarily, owing to extremely high pressures of the hot gases in the combustion chamber, is blocked by the check-valve.

During the downward movement of the plunger, the piston is rising on its compression stroke, forcing hot air up through the injection holes into the cup cavity in the tip of the nozzle. This emulsifies the fuel just inside the nozzle so, when injection begins, only the emulsion initially passes through the holes, giving a pilot injection effect, which reduces both the ignition delay and combustion noise. The duration of the pilot injection is reduced as the quantity of fuel metered into the nozzle increases. Therefore the timing of the start of main injection, which is when only liquid fuel is delivered through the nozzle holes, is automatically advanced as the load increases.

As the plunger is forced down again, it first covers the lower radial hole, trapping the metered quantity of fuel in the cup below, and opens the upper one so that circulation to the tank can begin again. From this point on, the pressure in the nozzle rises steeply to a maximum of up to 1310 bar, forcing the fuel out through tiny holes in its tip. At the same time, because the lower radial hole is covered, there is no possibility of any more fuel being delivered to the nozzle, so subsequently there is neither dribble nor after-injection. On completion of injection, the tapered end of the plunger remains for an instant on its seat in the nozzle until, under the influence of the spring at the upper end of the plunger, the plunger is lifted and the next upward stroke begins.

Chapter 6

Distributor type injection pumps

Distributor type pumps are defined as those in which one, two or more plungers serve all the cylinders of the engine, the fuel to be injected being directed to each cylinder in turn by a distributor. Two main types are currently in production: these are the rotary and axial plunger distributor pumps. In the rotary distributor type pump, sometimes referred to as simply the rotary type, the plungers reciprocate radially in the head of the distributor rotor while revolving around it axis, as in the Lucas and Stanadyne units. In the axial plunger type distributor pump, the pumping plunger is coaxial with the drive shaft and, rotating with it, serves also as the distributor, as in the Bosch VE Series.

Distributor pumps are generally lighter and more compact, and therefore less costly, than their in-line equivalents. Moreover, their governors and injection advance devices are simpler and smaller. This type of pump usually embodies also a transfer pump used not only for keeping it full of fuel but also for powering many of the control functions. Consequently, transfer pump delivery pressures of up to about 8 bar are needed. For this reason, vane type pumps are generally employed.

6.1 The Lucas DP series distributor pumps

Lucas Diesel Systems produce three types of distributor pump, DPA, DPS and the DPC. They are derived from the original Vernon Roosa design, which was first developed by the Hartford Machine Screw Company in the USA. All are flange mounted, and lubricated by the diesel fuel, which totally fills the cambox, where the pressure is maintained constantly above that of atmospheric.

The DPA, the first to be introduced, was originally intended for all applications. Subsequently, demands arose for pumps designed for specific applications. For instance, for tractors the requirement was for simple, robust low cost units. Cars, on the other hand needed more sophisticated controls while, for commercial vehicles, the need was for higher rates of injection and provision for features such as boost control.

Consequently, the DPA was followed by the DPC, which was designed specifically for indirect injection (IDI) engines, mainly installed in cars and car-derived vans. The DPS range, similar to the DPA, was introduced to

DISTRIBUTOR TYPE INJECTION PUMPS

meet the requirements for both high speed direct injection (HSDI) engines having piston-swept volumes of about 0.5 litres per cylinder, and direct injection engines of around 1 litre per cylinder, with or without turbocharging, for agricultural, industrial and light duty truck applications.

6.2 The DPA pump

The DPA pump is illustrated in Fig. 6.1, and shown diagrammatically in Fig. 6.2. Immediately inside one end, on the drive shaft, is the governor. An articulated splined muff coupling interconnects the drive shaft and the distributor rotor, in which are the diametrically opposed pumping plungers. An extension of the rotor forms a shaft, on the far end of which is the vane type transfer pump. This whole rotor assembly is housed in the steel *hydraulic head*, in which a ported distributor sleeve carries the section between the rotor and the transfer pump. The hydraulic head is spigoted into the cambox, the top of which is closed by an upturned bath tub shape cover accommodating the governor control springs and linkage. Driver-actuated control levers on vertical spindles pivot in bearings in the top of this cover.

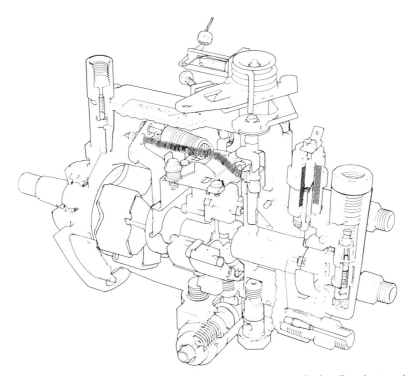

Fig. 6.1 The Lucas DPA distributor type injection pump was originally for all applications but this, the DPC version, was designed specifically to suit IDI engines

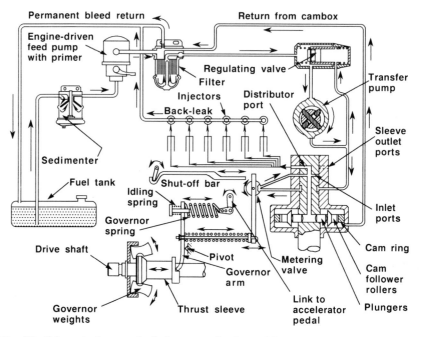

Fig. 6.2 Schematic diagram of a fuel system with a Lucas DP type injection pump

On their lower ends, these spindles carry levers linked to the governor controls.

The twin opposed plungers serving all cylinders reciprocate in a diametral bore in the injection pump rotor. They are pushed outwards by transfer pump pressure, to fill the pump chambers, and inwards by cams in a cam ring, to deliver the fuel through the distributor to the injectors. The cam followers are rollers carried in shoes sliding in radial slots in the rotor.

Outward travel of the plungers is determined by the quantity of fuel admitted, by a variable restrictor termed the *metering valve*, into the space between them. Their maximum travel is positively limited by lugs, extending one from each side of each shoe, registering in cam shape slots in two side-plates Fig. 6.3. Screws passed through slotted holes in these plates clamp them to the rotor. To adjust the maximum delivery, these screws are loosened, the plates rotated relative to the rotor so that the shoes ride up or down in the cam shape slots, and the screws tightened again. This adjustment of course is normally made only in the factory.

The delivery pressure of the vane type transfer pump, increases progressively with speed up to a maximum value determined by the setting of a pressure limiting valve housed in the end-plate next to the pump. As the rotor turns, the delivery port in the distributor sleeve is closed and an inlet port opened. The incoming fuel passes through the hole or, in some instances holes, in the hydraulic head and sleeve into a radial hole in the

DISTRIBUTOR TYPE INJECTION PUMPS

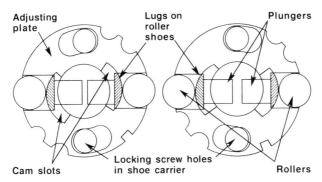

Fig.6.3 When the locking screws are loose, the twin adjusting plates can be rotated relative to the rotor to cause the cam section lugs on the roller follower shoes to ride up or down the cam slots, to adjust the maximum delivery of the pump. This of course is done only in the factory

rotor and then on through an axial hole to the space between the pump plungers. Although there may be only a single inlet port in the distributor sleeve, there are as many radial inlet holes in the rotor as cylinders on the engine. The pressure delivered by the transfer pump forces the plungers outwards, thus filling the space between them.

The rotary metering valve, which can be seen in Figs. 6.2, 6.7 and 6.8, is actuated by the governor Section 6.3, to regulate the fuelling to keep the speed constant regardless of variations in load. Because the quantity of fuel entering the space between the plungers increases progressively from idling to maximum power, the roller followers contact the cam profiles at points dependent upon accelerator pedal angle and governor linkage movement. To give a sharp cut-off of pressure at the end of injection and thus avoid dribbling from the nozzles, the cam lobe contours incorporate what are termed *retraction platforms*, Fig. 6.19, which allow the plungers to return a very short distance outwards before the fuel begins to enter again as they continue on their normal outward stroke. This instantly lowers the pressure at the nozzle, thus ensuring complete closure of the valve, while still maintaining a residual pressure in the line between injection phases.

When the plungers are driven inwards by the cams, the high pressure thus generated is transmitted along the axial hole in the extension of the rotor, where it communicates with a single radial delivery hole. As the distributor rotor turns, this hole aligns at regular timed intervals with ports in the distributor sleeve, whence the fuel is passed through ducts in the hydraulic head, through the delivery valves into the pipes serving each injector in turn. In other words, the porting arrangement for delivery is the inverse of that for ducting the incoming fuel to the plungers. Precise equality of timing intervals between injections is ensured by accuracy of spacing of the cams and delivery ports. Fuel intake and delivery conditions are illustrated in Fig. 6.4.

With this type of injection pump, there is a tendency for the timing to be retarded automatically as the load is reduced. This is because, owing to the

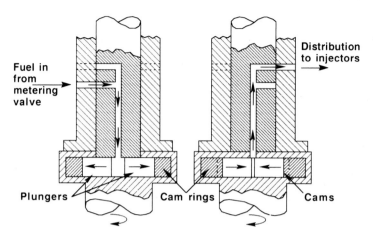

Fig. 6.4 Diagrammatic illustration showing the functioning of the distributor rotor, cam ring and plunger assembly

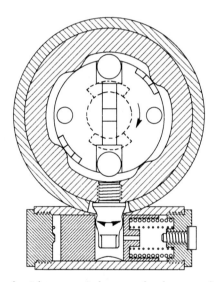

Fig. 6.5 The spring on the right progressively moves the plunger to the left, as the speed and therefore the pressure delivered from the transfer pump into the chamber on the left falls. This progressively retards the injection timing

reducing stroke of the pumping plungers, they contact the cams nearer to their peaks, and therefore later. Consequently, the basic timing of the start of injection is set for full load at maximum speed, and allowed to retard naturally as the speed and, incidentally with it, the pressure in the hydraulic advance/retard mechanism which rotates the cam ring, Fig. 6.5, falls. This mechanism, which is used also for retarding injection for starting, is as follows.

Screwed into the periphery of the cam ring is a ball-ended plug which projects radially outwards into a diametral hole in the piston in a hydraulic servo, the axis of which is aligned transversely relative to that of the cam ring. Axial displacement of the piston therefore rotates the cam ring. The piston is pushed towards one end of its cylinder by a coil spring, to retard the injection, and towards the other end, by fuel transfer pump pressure, to advance it. A non-return ball valve in the fuel delivery to the piston prevents the cam ring from being retarded by the impact of the roller followers on its cams.

For starting, or at low speed, when the transfer pressure is low, the spring will have moved the piston back to the retarded position. However, on some high speed indirect injection engines, the degree of retardation experienced by the previously mentioned late contact of the plungers with the cams, causes misfiring under light load. Consequently, a light load advance device acts on the metering valve, as described later, to increase the fuel pressure applied to the piston.

6.3 Governing

For the DPA pump, the all-speed governing system is based on the same principles as those for in-line pumps. The angle at which the accelerator pedal is held determines the position of the governor linkage at which the relevant spring balances the centrifugal force on the governor weights, to keep the engine speed constant regardless of variations in load.

Hydraulic governing used to be an option on this pump but, owing to a lack of demand, is no longer produced. Illustrated diagrammatically in Fig. 6.2 is the mechanically governed system for the DPA pump, while the hydraulic governor is depicted in Fig. 6.6. Both actuate the rotary valve that

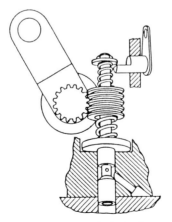

Fig. 6.6 Illustrating the principle of operation of the hydraulic governor that used to be available for the DP pump. Transfer pressure lifts the valve against the load exerted by the springs. The large disc serves as a damper

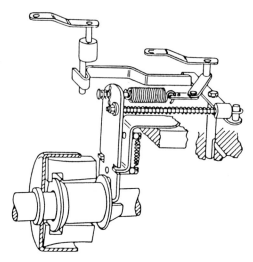

Fig. 6.7 Arrangement of the linkage between the mechanical governor, bottom left, and the metering valve, bottom right

meters the rate of flow of fuel to the pumping element. However, whereas with the mechanical governor displacement of the weights actuates a linkage connected to a lever on the upper end of a rotary valve, Fig. 6.7, with the so called hydraulic governing, the metering valve is lifted by hydraulic pressure acting against the force exerted by its return springs, as shown diagrammatically in Fig. 6.6. Around the spool valve stem, and free to slide axially along it is the rack, which is interposed between the main governor spring and a lighter idling spring. A large disc on the stem serves as a damper to prevent the valve from moving too precipitately and overshooting if the accelerator pedal is suddenly released.

During idling, the transfer pump delivery pressure, acting on the lower end of the spool valve, compresses the idling spring and keeps the idling speed constant. As the driver calls for increasing power output, the pump pressure, acting now against the main governor spring, progressively opens the valve and thus increases the rate of fuelling. As can be seen in the illustration, rotation of a small shut-down lever will cause an eccentric projection from its end to pull the valve spool back to shut off the fuel supply completely and stop the engine.

The mechanical governor is more precise but also larger and more costly. As the governor weights are displaced radially, they pivot in a star shape housing, in a manner similar to that of the C type governor, described in the second paragraph of Section 5.18. Their pivoting motion slides a sleeve along the rotor shaft. The opposite end of this sleeve bears against the lower end of a lever pivoted on a knife edge, so that its upper end will actuate the linkage that rotates the metering valve stem. As can be seen from Figs. 6.2 and 6.7, this valve regulates the rate at which the fuel flowing from the transfer pump is delivered through a groove parallel to the axis of this valve, into ducts leading to the space between the opposed plungers.

Mounted externally on the top cover are two levers, one is linked to the accelerator pedal and the other to a shut-down control. The upper of the two springs that can be seen in the illustrations is the main governor spring and is in tension, except for idling. During idling, the shorter spring, which is in compression on the other side of the vertical lever, comes into operation. These springs provide the balancing force against which the governor operates to maintain constant speed regardless of changes in load.

The second lever is for stopping the engine. Inside the cover, an eccentric peg at the lower end of this lever registers between the arms of a U formed in one end of a strut. The opposite end of the strut bears against against one end of a lever fixed approximately mid-way between its ends on the metering valve. When this control is actuated, the eccentric peg moves the strut axially, causing it to push the end of the lever on the metering valve, rotating it far enough to shut off the fuel supply. Connected to the other end of this lever is a rod, the far end of which is free to slide axially through a hole in the vertical lever. Around the rod is a lightly loaded compression spring, which does not have any effect on the governing. Its function is simply to return the control rack to the idling position after the fuel shut-off control has been released.

6.4 The Lucas DPS pump, with torque and boost control

To provide for excess fuel and both torque and boost control for four- and six-cylinder engines, Lucas introduced the DPS pump, Fig. 6.8. As regards overall principles of operation, it is the same as the DPA. However, it can be supplied with either one or two pairs of diametrically opposed plungers, with their axes in a common plane. The four plunger version is for larger and more powerful engines, including those of the 90 deg and 60 deg V layouts. Provision can be made for belt drive, if required. Both two-speed (maximum and idling speeds) and all-speed governing is available.

This pump has a number of other features, including externally adjustable maximum fuel delivery, torque or boost control for turbocharged engines, automatic excess fuel for starting, and electrical (key switch) shutdown. The system is illustrated diagrammatically in Fig. 6.9.

This pump has a stiffer drive than the DPA. For taking the radial loading imposed by a belt drive, two bearings, 18 and 20 in Fig. 6.8, one each side of the governor assembly, carry the shaft. A tongue on the end of the rotor registers in a slot in the end of the drive shaft, to transmit the drive between the two. The pairs of plungers reciprocate in one or, if there are four plungers, two diametral bores in the rotor.

Both the filling and distribution porting arrangements in the rotor, Fig. 6.10, are similar to those of the DPA pump. The end of the hydraulic head remote from the rotor is counterbored to receive the eccentric cam form, transfer pump liner, Fig, 6.11. If there is no external lift pump, the pressure

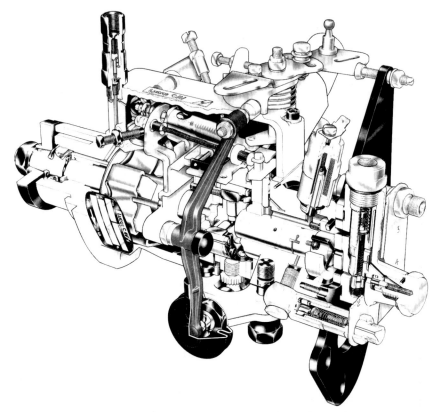

Fig. 6.8 This version of the Lucas DPS distributor type pump, designed for a heavy duty belt drive, is for high speed DI engines. Although similar to the DPA, it has a stiffer drive

in the feed line is generally sub-atmospheric, so there may be an orifice in the top of this counterbore, bottom left in Fig. 6.11. This is for venting to the housing any air that might get into the supply, so that it will not be drawn in to the pumping elements.

Some rotors have on the same plane as the distributor port, an equalising groove around most of the periphery of the rotor, as also shown in Fig. 6.11. This, by interconnecting all the delivery lines to the injectors except that which is about to function, balances their residual pressures.

A high pressure delivery valve is screwed into a banjo connection from each delivery port around the pump. It functions on the principle already described, in connection with the in-line pumps, in Sections 5.10 and 12. Small holes drilled axially through it allow fuel to flow back, when injection terminates, into the equalising slot around the distributor rotor.

For key starting and stopping the engine, a solenoid-actuated valve is screwed into the top of the hydraulic head. When closed, this valve shuts off the supply of fuel from the transfer pump to the metering valve. The

DISTRIBUTOR TYPE INJECTION PUMPS

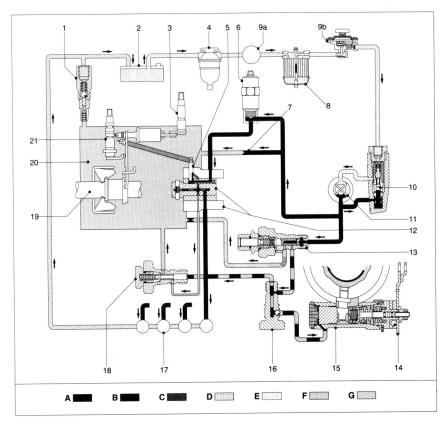

Fig. 6.9 Recommended layout of fuel system for the Lucas DPS injection pump
A Injection pressure, B Transfer pump pressure, C Metering pressure, D Differential pressure, E Feed pressure, F Cambox pressure, G Back leakage
1 Pressurising valve, 2 Fuel tank, 3 Throttle shaft, 4 Sedimenter or water stop, 5 Metering valve, 6 Shut-off solenoid, 7 Vent orifice, 8 Filter, 9 (a) Feed pump, when fitted, (b) hand primer, when fitted, 10 Regulating valve, 11 Transfer pump, 12 Hydraulic head and rotor, 13 Latch valve, 14 Manual idle advance lever, 15 Automatic advance and retard unit, 16 Head locating fitting, 17 Injector, 18 Rotor vent switch valve, 19 Two-speed mechanical governor and control linkage, 20 Cam box, 21 Idle shaft

solenoid is energised, to open the valve, when the key is turned for starting the engine, and does not close again until the engine is switched off.

6.5 DPS fuel supply system

From Fig. 6.9, it can be seen that fuel is drawn by the transfer pump from the tank, through a sedimenter and filter, for delivery to the regulating valve 10. In some installations, a feed pump may be needed in the supply line between the sedimenter and filter. Alternatively, there may be, after the filter,

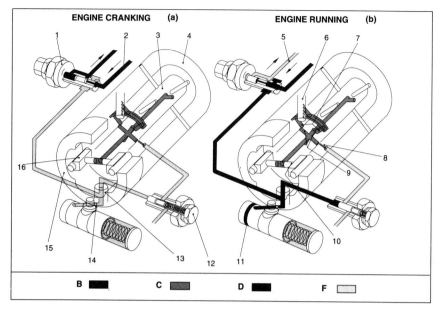

Fig. 6.10 Diagram showing the layout of the system and distribution of the fuel to the latch and rotor switch vent valves (a) at cranking speeds and (b) when the engine fires and runs under its own power
B Transfer pressure, C Metering pressure, D Differential pressure, F Cambox pressure
1 Latch valve, 2 Inlet from transfer pump, 3 Distributor rotor, 4 Hydraulic head, 5 Return to cambox, 6 Metering valve, 7 Filling ports – hydraulic head, 8 Vent orifice, 9 Rotor inlet ports, 10 Pump plunger, 11 Pressure chamber auto-advance unit, 12 Rotor vent switch valve 13 Head locating fitting 14 Ball valve 15 Cam ring 16 Roller and shoe

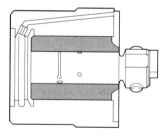

Fig. 6.11 Section through the hydraulic head. Right to left: plunger assembly, rotor filling ports and distributor port and peripheral grove

a manually actuated priming pump for priming the system, for example, following a filter element change or simply for starting after the tank has been emptied.

Fuel from the tank enters the pump through the upper end of the regulating valve, Fig. 6.12. Here, it passes through the fine mesh filter sleeve, bypassing the spring-loaded valve and going directly to the inlet side of the transfer pump. Similarly, fuel delivered from the manual priming pump when the engine is stationary also bypasses the transfer pump. The output from transfer

DISTRIBUTOR TYPE INJECTION PUMPS

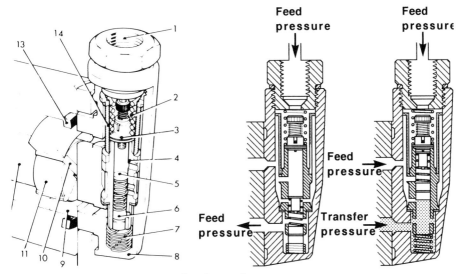

Fig. 6.12 Regulating valve mounted on the transfer pump
1 Fuel inlet, 2 Retaining spring, 3 Transfer pressure adjuster, 4 Regulating sleeve, 5 Peg and spring, 6 Regulating piston, 7 Priming spring, 8 End plate, 9 Eccentric liner, 10 Pump blades, 11 Transfer pump rotor, 12 Distributor rotor, 13 Rubber sealing ring, 14 Coarse Nylon filter

pump passes back into the valve which automatically regulates the output pressure, causing it to build up progressively with engine speed.

How the valve functions can be seen from the illustration. There are four radial ports, one above the other, in the sleeve in which the regulating piston moves. Of these four, the lowest is in fact a diametrically opposed pair, termed *priming ports*. When the engine is started, fuel from the transfer pump enters through the lowest port, lifting the piston 6, until the regulating spring 5 contacts the adjustable stop 3 above. As the engine speed increases, the regulating spring 5 is compressed, allowing the piston to rise and uncover an increasing area of a regulating port above, which bypasses fuel upwards to the inlet side of the pump. The rate at which the transfer pressure rises is adjusted, to suit each engine application, by means of the *transfer pressure adjuster screw* 3.

If the manual primer is used prior to starting, its passage to the injection pump is blocked by the stationary transfer pump. Consequently, the priming pressure forces the regulating piston down, compressing the priming spring 7 and uncovering the priming ports through which the fuel passes to the delivery side of the transfer pump. From here, it goes through two passages, one to the injection pump and the other to the latch valve, 13 in Fig. 6.9.

Under normal running conditions, the output from the transfer pump is also delivered to these two passages. However, as will be explained later, the latch valve 13 is closed except under starting conditions, leaving only the passage to the injection pump open. Therefore, flowing past the duct to the vent orifice 7, the fuel goes on through the solenoid-actuated shut-off valve 6 to a hole drilled from the top of the hydraulic head sleeve into annular

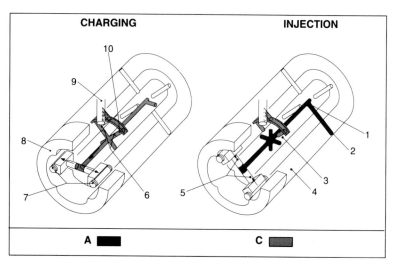

Fig. 6.13 Diagram illustrating (a) fuel charging and (b) injection cycles
A Injection pressure, C Metering pressure
1 Single outlet fuel delivery port, 2 Outlet port in hydraulic head, 3 Rotor, 4 Hydraulic head, 5 Pump plungers, 6 Charging ports in rotor, 7 Rollers, 8 Cam ring, 9 Metering valve bore, 10 Filling ports in hydraulic head

grooves around the hydraulic head, whence it flows through the metering valve 5. This valve operates on the same principle as that of the DPA pump, described in Section 6.3 and illustrated in Fig. 6.7.

The arrangement of the holes in the rotor, Fig. 6.10, is similar to that of their counterparts in the DPA pump, as also is the delivery of of fuel at transfer pressure to the metering valve and, at metered pressure, to the injection pump. Finally, the fuel is delivered at high pressure to the distributor ports. All this is as described in Section 6.2. Also as in the DPA, a controlled degree of leakage passes from the hydraulic head into the cam box, where the maximum pressure is limited by the pressurising valve 1, Fig 6.9, which is illustrated in more detail in Fig. 6.14.

6.6 Engine starting — DPS latch and rotor vent valves

In addition to performing their primary functions, the latch valve and the rotor vent switch valve, 1 and 12 respectively in Fig. 6.10 and 13 and 18 in Fig. 6.9, retain the hydraulic head in the pump body. At cranking speeds, the latch valve is closed to prevent transfer pressure from reaching the automatic injection advance unit and rotor vent switch valve, 15 and 18 respectively in Fig. 6.9. This prevents the injection from being advanced until the engine starts to run properly. When it does so, the increasing pressure lifts the latch valve, allowing fuel at transfer pressure to flow through to the pressure chamber in the advance unit.

DISTRIBUTOR TYPE INJECTION PUMPS

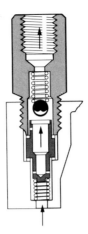

Fig. 6.14 The Lucas DPS pressurising valve

At cranking speeds, which is when the rotor self-venting feature is in operation and the latch valve is closed, the rotor vent switch valve is held open by its spring. In this condition, the small vent orifice 8 in the hydraulic head sleeve, Fig. 6.10, communicates in turn with each of the rotor inlet ports 9. Consequently, any air that might be trapped in the pumping elements is forced, by the residual pumping pressure (from the previous pumping cycle) through this hole, but only during the first 12 deg of rotation past the point where venting began. For the next 11 deg of rotation, the vent orifice is in communication with the two oblique filling ports 7 in the hydraulic head and air is again forced, but this time at metering pressure, through the orifice. In both instances, a mixture of air and fuel, passes through a passage in the hydraulic head, across the rotor vent switch valve, 12 in Fig. 6.10, into the cam box, whence it is vented, through the pressurising valve to the tank.

After the engine has fired and begins to run normally, the build up of transfer pressure lifts the latch valve 1 against its return spring. This, allows fuel under transfer pressure to pass through the latch valve to the rotor vent switch valve, 18 in Fig. 6.9, lifting it against its return spring and thus closing the rotor vent passage.

6.7 Limiting maximum fuel delivery

Whereas maximum fuel delivery in the DPA is fixed by the setting of cam plates secured by screws to the rotor, in the DPS it can be varied by rotating what are termed *scroll plates*, again assembled one each side of the cam ring, 9 in Fig. 6.15. The maximum outward travel of the rollers is limited by the spiral profiles 6 of the inner edges of these two plates, but the plates are free to rotate though only a limited angle, which can be adjusted by means of a screw and lock nut on the casing, 10 in Fig. 6.15.

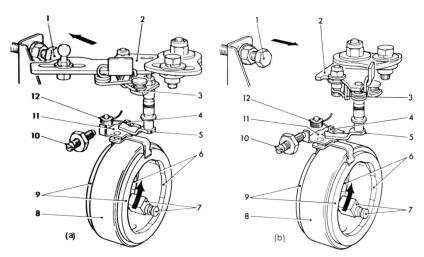

Fig. 6.15 In the DPS pump, the maximum fuel delivery is controlled by scroll plates shown here at (a) in the excess fuel position with the accelerator pedal released, and at (b) in the maximum fuel position with the accelerator pedal depressed
1 Anti-stall stop, 2 lever connection to accelerator pedal, 3 Excess fuel linkage pin, 4 Inner tongue on link plate, 5 excess fuel spindle and lever, 6 Scroll plate profiles, 7 Roller and shoes, 8 Cam ring, 9 Scroll plates, 10 Maximum fuel adjustment screw, 11 Link plate, 12 Link plate spring

These plates are also used for providing excess fuel for cold starting. For this purpose, when the accelerator pedal is in the idling position, a lever 5 actuated by its linkage to the governor pulls lug 4 on the scroll plate link 11 back to the excess fuel position. When the engine fires and runs steadily at speeds above that for cranking, excess fuel is terminated by the action of the governor on the metering valve, which reduces the rate of fuelling to that required for idling. Another action that is initiated as soon as the engine is running, is retardation of the injection timing by the automatic advance and retard unit, Fig. 6.16. Subsequently, as the engine speed rises further, it is progressively advanced.

In later models, this system has been replaced by a hydraulically actuated device, Fig. 6.17. To obtain excess fuel for starting, a spring-loaded hydraulic piston moves the scroll plate. The mechanism functions as follows.

When the engine is stopped, and the transfer pressure therefore falls, the piston is moved by its return spring to the right, as viewed in the illustration. This leaves it in the excess fuel position, ready for re-starting.

As the engine is cranked, fuel under transfer pressure which, under these conditions is low, passes through orifices A and B into the cambox. The restrictions offered differentially by these two orifices are such that the fuel between them is at an intermediate pressure. The excess fuel control latch valve is therefore kept closed by the pre-load in its return spring, supplemented by this intermediate pressure.

When the engine starts, the rising transfer pressure pushes this valve upwards to the right, against the influence of its return spring, closing port

DISTRIBUTOR TYPE INJECTION PUMPS

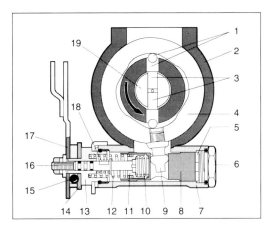

Fig. 6.16 The DPS automatic advance and start-retard unit. Automatic retard comes into operation while the engine is being started and disengages when it is running
1 Roller and shoe, 2 Pump housing, 3 Plungers, 4 Cam ring, 5 Transfer pressure chamber, 6 Plug at pressure end, 7 Auto-advance housing, 8 Piston, 9 Cam advance screw, 10 First stage (retard) spring, 11 Spring plunger, 12 Advance spring, 13 Advance spring end cap, 14 Detent plate, 15 Balls, 16 Spindle, 17 Manual cold idle advance lever, 18 Spindle spring, 19 Distributor rotor

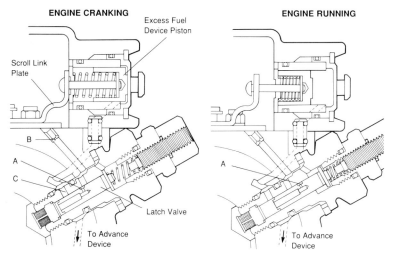

Fig. 6.17 More recently, this hydraulic excess fuel unit has become available. The orifices and portrs A,B and C are referred to in the text

A and uncovering the port C through which fuel under transfer pressure passes, to push the excess fuel piston to the left, against its return spring. This moves the fuelling stop to its normal maximum fuel position. The uncovering of the port C also opens the way for fuel under transfer pressure to pass to the advance device, which of course must not operate during starting.

6.8 The two-speed governor

Three, four or six weights can be carried in the star shape housing. The axially sliding sleeve 8 and vertical lever 7, in Fig. 6.18, rotate the metering valve through linkages similar to those of the DPA pump. Not shown in the illustration is an external lever arm secured to the top of the idling lever shaft 12, which is rotated into the position in the illustration by the accelerator pedal when it is released to the idling position. The end of lever 11 carries a stop against which the free end of the idling leaf-spring 6 bears. Consequently, at idling speeds, the degree of rotation of the metering valve is determined by the balance between the forces exerted on the vertical lever 7 by the idle spring 6 and by the governor weights on the axially sliding sleeve.

At intermediate speeds, the pre-load in the main governor spring 14 is such that it is not deflected, so the linkage between the accelerator pedal and metering valve is, in effect, solid. Consequently, the angle of rotation of the metering valve is a function solely of pedal movement. As the predetermined, or governed, maximum engine speed is approached, the centrifugal force on the governor weights increases to the point at which, acting through the spring and lever, it compresses the governor spring. In this condition, a decrease in load will not increase the speed though, by further depression of the accelerator pedal, the compression in the governor spring, and therefore the rate of fuelling, can be increased up to the maximum rated torque, which is fixed by an externally adjustable stop. The speed of course remains dependent on the load, and governed so that it remains constant for any given pedal angle.

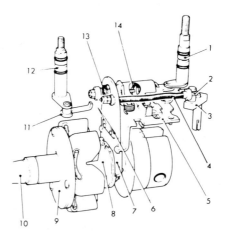

Fig. 6.18 The two-speed governor for the DPS pump is claimed to give rapid response to pedal movement and driving characteristics comparable to those obtained with a spark ingnition engine
1 Shaft for accelerator pedal controlled lever, 2 Linkage hook, 3 Metering valve, 4 Governor link and control spring, 5 Control bracket, 6 Idling leaf spring, 7 Governor arm, 8 Thrust sleeve, 9 Governor flyweight assembly, 10 Driveshaft, 11 Idle actuator, 12 Idling lever spindle, 13 Anti-stall device, 14 Main governor spring

DISTRIBUTOR TYPE INJECTION PUMPS

During rapid deceleration from high speed, the main governor spring may be released more rapidly than the speed falls. In these circumstances, there would be an imbalance between the force exerted by the governor weights and that applied to the upper end of the vertical lever. The outcome could be that the metering valve would be rotated past its delivery shut-off position and cause the engine to stall. To avoid this, an anti-stall spring 13 is interposed between the anchorage for the main governor spring and the vertical lever 7. When it is compressed, it opens the metering valve again rapidly enough to prevent stalling.

6.9 Scroll plates and boost control

As mentioned previously, the scroll plates, one each side of the cam ring, serve two purposes. First, they limit the outward movement of the plungers and therefore the maximum fuelling. Secondly, they can be used to provide excess fuel for starting.

From Fig. 6.15, it can be seen that the scroll plates are rotated by lugs extending down from the link plate 11 to register in slots in their top edges. The link plate slides tangentially relative to the scroll plates. It can be moved in one direction by rotation of the excess fuel shaft and levers, 3 and 5, and in the other direction as far as the stop 10, by the spring 12.

When the throttle is closed against the anti-stall stop 1, and the engine is rotating at cranking speed, the cam ring is in its retarded position. At the same time, the excess fuel lever 3, bearing against the lug 4, pulls the link plate back, against the direction of rotation of the pump. As it does so, it loads the spring 12 and rotates the scroll plates clockwise, as viewed in the illustration.

As the rotor filling port opens, metering pressure forces the plungers outwards until their rollers and shoes are stopped by the profiles 6 of the scroll plates at point (a) in the upper diagram in Fig. 6.19, which is further out than the normal maximum. They continue to move outwards against the scroll plate profiles until the filling port closes, at (b). Finally, they contact the cam at point (c), which moves them inwards again to deliver the excess fuel that the plungers have, in the meantime, drawn in. As soon as the engine is running normally, the excess fuel condition is cancelled by the governor, which rotates the the metering valve to reduce the rate of fuelling to that required for idling.

As the throttle is opened further, the lever 5 is moved away from the lug 4, allowing the spring to return the link plate on to the maximum fuelling stop 10. This rotates the scroll plates in the direction of rotation of the pump, to the point at which their profiles limit the outward motion of the rollers and shoes to the normal maximum. In this situation, illustrated in the lower diagram, the rollers contact the scroll plates at (d) as the filling port opens, and follow the profile of the scroll to (e). When the delivery port opens at (f), they are thrust inwards again by the cam.

For turbocharged engines a boost control unit can be fitted to increase

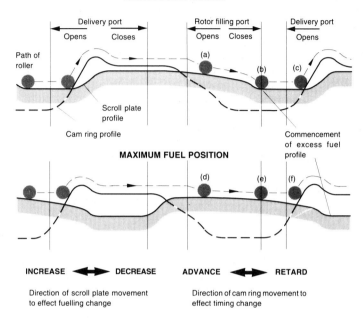

Fig. 6.19 Diagram showing how the scroll plate movement controlls the paths the rollers take on approaching and leaving the cams

the fuelling with boost pressure. It is an actuator in which one side of a diaphragm is subjected to boost pressure which, on the other, is balanced by the force exerted by a coil spring. The diaphragm moves a push rod, which slides the link plate tangentially relative to the scroll plates, thus rotating them to increase or decrease the rate of fuelling.

6.10 Automatic advance and retard unit

This unit automatically advances injection as engine speed increases. It can also incorporate an automatic retard system that comes into operation when the engine is started and then, when it is running, cuts out again. A manual advance device can be added too, so that the driver can maintain idling stability in very cold conditions.

The unit is illustrated in Fig. 6.16. A spigot screwed into the cam ring registers in a diametral hole through the piston 8 which, sliding in the bore in the advance housing 7, rotates the cam ring in the pump housing, to advance or retard the injection. Fuel flows, at advance pressure, from the latch valve, through the head location fitting, into the chamber 5. The pressure pushes the piston to the left, as viewed in the illustration, until the force it exerts is balanced by compression of the advance spring 12 in the chamber at the opposite end.

DISTRIBUTOR TYPE INJECTION PUMPS

Start retard is controlled by the low rated first stage (or retard) spring 10 accommodated in a counterbore in the piston and compressed between it and the seat 11 for the advance spring. This seat is mounted on the end of the spindle 16 that is free to slide axially in the end cap 13, on which the other end of the advance spring seats. Compressed between the end cap and a collar on the spindle is the spindle spring 18 the function of which is to load the three ball detents 15 for the manual advance lever 17.

When the engine is stopped and the latch valve therefore closed, there is no fuel pressure to act on the end of the piston so, acted upon by both the advance and first stage springs, it rests against the pressure plug 6 and retards the injection. As soon as the engine is started, the transfer pressure moves the piston along the bore, compressing the first stage spring and advancing the injection to the idling position.

As the engine speed increases, the correspondingly rising pressure moves the piston further along the bore, compressing the advance spring as it goes. This rotates the cam ring in the direction opposite to that of rotation of the pump and thus advances the injection phase. Reductions in speed have the opposite effect, causing the piston to return.

The manual idling advance lever, for idling in very cold conditions, is cable-actuated by the driver. When it is rotated, the three balls roll out of their detents, thus moving the spindle, and with it the seat 11 for the advance spring, axially outwards. Consequently, when the engine starts and the latch valve has opened to subject the piston to the advance pressure, it can move beyond the normal advance position for idling.

Helping to retain the advance unit, and locating the hydraulic head relative to the pump housing, is the head locating fitting, or damper assembly, Fig. 6.20. This also serves to connect the fuel passages in the hydraulic head with the advance unit. The ball-valve 5 hydraulically locks the advance and cam ring mechanism which, otherwise, would be deflected in the retard direction

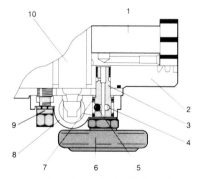

Fig. 6.20   The damper unit serves also as both a hydraulic head location fitting and for connecting the fuel passages in the hydraulic head with the injection advance unit
1 Distributor rotor, 2 Hydraulic head, 3 Head location fitting, 4 Bypass hole, 5 ball valve, 6 Damper assembly, 7 O-ring seals, 8 Automatic start-advance and retard unit, 9 Dome nut, 10 Cam ring

by the impacts of the rollers on the cams. When the valve is closed, fluid flow can still occur but is restricted by the bypass hole 4. The damper assembly 6 is fitted only in some installations in which pulses or other disturbances in the transfer pressure cause inconsistency of rotor filling.

6.11 The Lucas DPC pump

This pump, Fig. 6.21, has been designed primarily for indirect injection engines up to 2.5 litres capacity for cars. It has two pumping plungers housed in a diametral bore in the rotor. In most other respects it is similar to the DPA and DPS, so only the significant differences will be described here.

A two-speed governor is normally installed in it, though an all-speed version is available. The linkage between the governor and metering valve is similar, but a different type of anti-stall device is employed. When the accelerator pedal is suddenly released from high speed, the governor could tend to overshoot, moving the metering valve beyond its idling position and stalling the engine. To ensure that it cannot do so, an anti-stall screw stop 1, in Fig. 6.22, is fitted beyond the idling position. The resulting extra pressure on the idling spring leaf spring 4 returns the governor arm to the idling position.

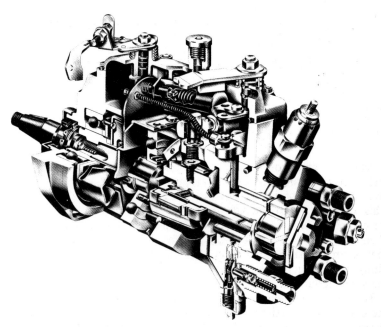

Fig. 6.21 The Lucas DPC pump was designed primarily for indirect injection engines of up to 2.5 litres for cars

DISTRIBUTOR TYPE INJECTION PUMPS

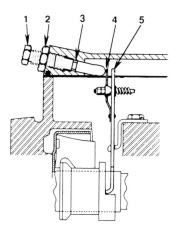

Fig. 6.22 This is the anti stall device on the Lucas DPC pump
1 Anti-stall screw, 2 Lock nut, 3 Seal, 4 Leaf spring, 5 Governor arm

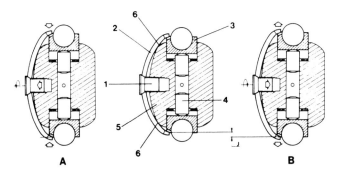

Fig. 6.23 Maximum fuel adjustment device
1 Adjustment screw, 2 Fuel adjustment plate, 3 Shoe and roller assemblies, 4 Plungers, 5 Rotor, 6 Points of contact between plate and rotor

As can be seen from Fig. 6.23, the provision for adjustment of maximum fuel delivery is different too. A spring steel strip 2, termed the *maximum fuel adjusting plate*, is held against the rotor head by a self locking screw 1. This forces the strip against the rotor 5 at the two contact lines 6. Access for adjusting the self locking screw is gained through a hole in the cam ring. Tightening the screw moves the ends of the plate further apart and, since they form the stops that limit the maximum outward movement of the plungers, this increases the maximum fuel delivery. Conversely, loosening the screw reduces it.

To enable the fuelling to be increased for starting, slots are machined in the ends of the maximum fuel adjusting plate and in the shoulders of the shoes, as shown in Fig. 6.24. When the tongues thus formed in the plate are in line with the slots in the shoes, the outward movement of the plungers is

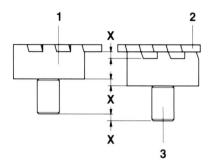

Fig. 6.24 To obtain excess fuel for starting, slots on the outer faces of the roller shoes (1) are moved into alignment with lugs on the maximum fuel adjustment plate (2) so that the shoes, and therefore the plungers (3), are free to move outward further outwards, as in the left-hand diagram

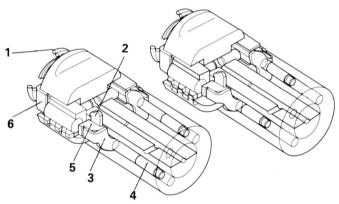

Fig. 6.25 Excess fuel actuation mechanism

increased by the amount X. As can be seen from Fig. 6.25, the rollers in their shoes are located in grooves machined longitudinally in the periphery of the rotor, where they are embraced by a carriage which is slid axially to move the slots into and out of engagement. The carriage assembly comprises four thrust pads 5, each pair being separated by a spacer 2 and held in position by two retaining plates 3 and 6. It is moved by the excess fuel delivery pistons 4 under the influence of transfer pump delivery pressure, and returned to the normal fuelling position by the return spring 1. A stop-washer in the end adjacent to the hydraulic head limits the motion of the pistons.

Excess fuel is controlled by the excess fuel delivery valve, Fig. 6.26, situated in a hollow screw that helps to secure the hydraulic head to the casing. This is a valve that is closed by a coil spring and opened by the rising pressure delivered by the transfer pump. In the absence of transfer pump

DISTRIBUTOR TYPE INJECTION PUMPS

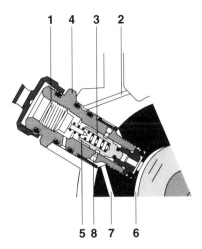

Fig. 6.26 Excess fuel delivery valve, closed by the spring (3) and opened by the pressure delivered by the transfer pump
1 Adjustment screw, 2 Back leakage to cam box, 3 Valve spring, 4 Vave body, 5 Valve, 6 Annular groove under transfer pressure, 7 Passage to excess fuel pistons, 8 Port communicating with valve

pressure, when the engine is stopped, the excess fuel delivery carriage is moved by spring 1 in Fig. 6.25, into the excess fuel position.

The transfer pressure, increasing with speed, passes through groove 6 in the hydraulic head and lifts the valve off its seat. This allows transfer pressure to pass through port 7, in Fig. 26, to both the injection advance device and excess fuel pistons in Fig. 6.25. These pistons move the excess fuel carriage, against its return spring, back to the position at which maximum fuel is obtainable. The valve spring seats against the end of a screw by means of which its pre-load can be adjusted to regulate the valve lift pressure.

For automatically retarding the timing of injection for starting and advancing it with increasing speed, a mechanism, Fig. 6.27, is used which is similar in principle but differs in detail from that of the DPS, Fig. 6.16. How it functions is obvious from the illustration and the description of that for the DPS pump, Section 6.7. except in that the automatic start-retard mechanism differs from that of the DPS as follows.

The first stage, or start-retard, spring 5 is interposed between the piston and the spring plate, or moving stop, 4. When the excess fuel valve opens, fuel at transfer pressure pushes the piston to the left, compressing the first stage spring until it contacts the spring plate. This is the zero advance position. Only when the transfer pressure is high enough to compress the main advance spring 3 will the injection timing begin to advance with engine speed. As in the DPS, a non-return valve prevents the pressure pulses due to the rollers contacting the cams from retarding the injection timing.

An optional addition to the automatic start-retard and speed advance mechanism is an automatic cold advance override, Fig. 6.28. As the engine becomes cold, the plate on the inner end of the spindle that carries the control

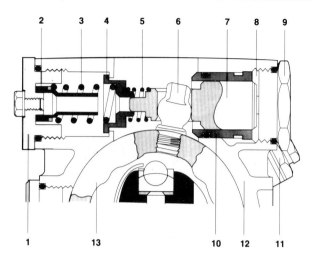

*Fig. 6.27 The DPC advance and retard mechanism differs slightly from that for the DPS pump
1 Plug for retaining spring, 2 Maximum advance stop, 3 Main advance spring, 4 Moving stop (spring plate), 5 Retard spring, 6 Cam screw, 7 Automatic advance piston, 8 Pressure chamber, 9 Plug (pressure end), 10 Automatic advance sleeve, 11 Head location fitting, 12 Pump housing, 13 Cam ring*

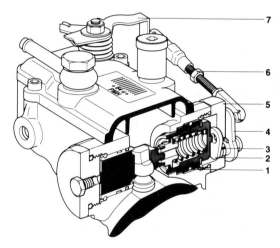

*Fig. 6.28 Automatic advance override. In the cold condition, the fast idling lever acts on the linkage to rotate the control shaft until the hole in the stop plate is in line with the spigot on the piston
1 Light load advance piston, 2 Stop plate, 3 Control shaft, 4 Control lever, 5 Plug, 6 Linkage, 7 Idling lever*

lever 4 is rotated either by a thermo-actuator or by a fast idle linkage until the hole in it is in line with the centrally positioned spigot on the end of the piston 1. This increases the stroke of the piston so that it can advance the cam ring further, the degree of advance override being determined by the

DISTRIBUTOR TYPE INJECTION PUMPS

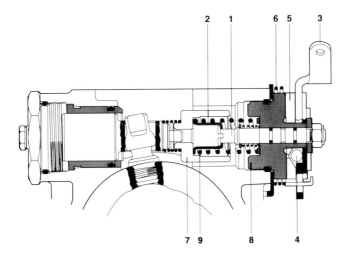

Fig. 6.29 Manual advance override for reducing smoke emission when the engine is idling in the cold condition
1 Shaft, 2 Cup, 3 Control lever, 4 Steel balls, 5 Plug, 6 Linkage, 7 Idling lever, 8 Advance plug, 9 Advance spring

length of the spigot. As the engine speed increases after starting, the excess fuel valve opens, allowing fuel at transfer pressure to move the advance piston, and therefore cam ring, beyond the normal idling advance position, terminating the excess fuel supply and start retard.

To cancel the advance override when the engine is warm, it is necessary only to depress the accelerator pedal. This causes the transfer pressure to move the light load advance piston 1 to the full load position, disengaging it from the plate on the control shaft. The control lever 4, no longer pulled back by the thermo-actuator, rotates the spindle 3 until the spigot and hole are no longer aligned. Consequently, when the accelerator pedal is returned, the piston stops when the spigot comes up against the plate.

An alternative manual light load advance mechanism is illustrated in Fig. 6.29. Movement of lever 3 rotates shaft 1, causing three steel balls 4 to ride out of their detents. This partially compresses the advance spring 9. When the excess fuel differential valve opens, transfer pressure moves the advance piston, and correspondingly rotating the cam ring, to the advance override position.

A correction mechanism termed the light-load advance is available too. Under light load conditions, when the fuel–air mixture is lean and injected late in the cycle, it superimposes upon the speed-controlled advance a further advance of the injection timing. The object is not only to compensate for the natural tendency to retard as load is reduced but also to provide additional advance without which it might be difficult to suppress smoke and noise under light load. How it works is illustrated diagrammatically in Fig. 6.30. In Fig. 6.31, the excess fuel valve and low-load advance valve 10 are shown on opposite sides of the section through the pumping element housing.

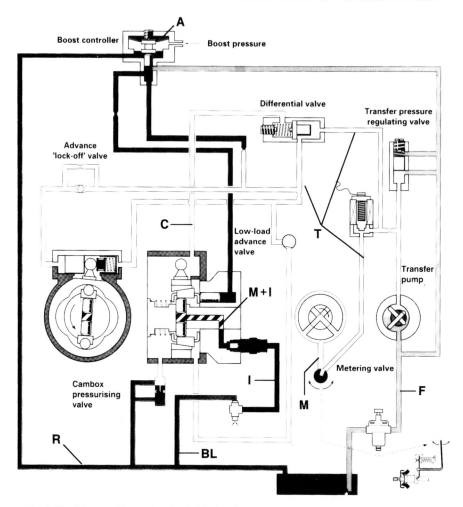

Fig. 6.30 Diagram illustrating the DPC distributor type pump system incorporating a boost controller. To the left of and below the transfer pressure regulating valve is a solenoid actuated low load delivery valve. The key letters indicate the different pressures in the system, as follows:
A Atmospheric, BL Back leakage, C Cambox, F Fuel feed, I Injection
M + I Alternately metering and injection, R Return to tank, T Transfer pressure

Hydraulic advance is another option, Fig. 6.31. Inclined groove G in the metering valve directs transfer pressure through the duct to the low load advance valve 10, so rotation of the metering valve varies the open area of the orifice at the inner end of the duct. Metering stop 14 and screw 15 are used, in conjunction with the washer beneath lever 11, to adjust the basic setting of the orifice.

At idling or fast idling and low load, valve 10 is open, as in Fig. 6.31. Orifice G is wide open so piston 5, subjected to only the low delivery pressure

DISTRIBUTOR TYPE INJECTION PUMPS

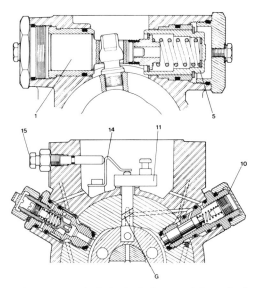

Fig. 6.31   Hydraulic advance overrride device in the low speed, heavy load condition

from the pump, is pushed by its spring to the right. Valve 10 is open also at low speed and heavy load, because the transfer pressure is still not high enough to close it but, with the metering valve in the full load position, variable orifice G is closed. The higher transfer pressure therefore moves the piston 5 to the left, against its full load stop, so only speed controlled advance piston is operating. Above a preset speed, increasing pump delivery pressure moves the piston in valve 10 up to the right so, as speed and load increase, piston 5 is pushed towards its full load stop, when the speed controlled advance piston takes over completely.

6.12 External control of low-load advance

As previously mentioned, low-load control advance can be effected by an external valve. In this case, all the controls are embodied in the low-load advance valve, leaving the metering valve to perform the sole function of metering the quantity of fuel to be injected. The arrangement and details of the valve assembly are as shown in Figs. 6.30, 33 and 34, and a solenoid actuated advance override mechanism is illustrated in Fig. 6.32.

In Fig. 6.33, shaft 2, rotating in the body of the low-load advance valve 9 and connected to the accelerator control by link 6, controls the variable valve orifice G. The fixed orifice between the transfer pressure regulating valve and the low load advance valve in Fig. 6.30 is actually adjacent to the passage B shown in Fig. 6.34. The amount by which the orifice G opens for any given accelerator pedal position can be adjusted By varying the length of the link 6 in Fig. 6.33. Adjustment of the transfer pressure at which the

146　　　　　　　　　　　　　　　　　　DISTRIBUTOR TYPE INJECTION PUMPS

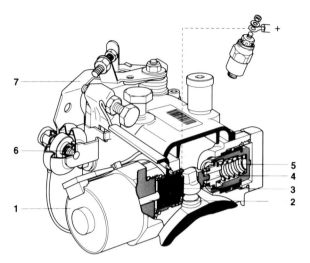

Fig. 6.32 Solenoid actuated advace override for the DPC pump
1 Advance override solenoid, 2 Plunger, 3 Advance piston, 4 Stop, 5 Advance spring, 6 Switch, 7 Idling lever

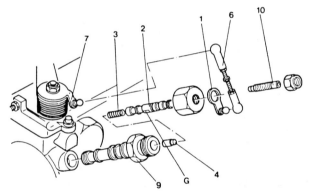

Fig. 6.33 Control mechanism for low load advance by means of an external valve, which leaves the metering valve with the sole function of metering the fuel to be injected
1 Actuation lever, 2 Shaft, 3 Spring for loading the low load advance piston, 4 Low load advance piston, 5 See Fig.6.32, 6 Adjustable connecting link, 7 Throttle lever, 8 Fixed orifice, 9 Valve body, 10 Spring load adjustment screw, G Variable orifice

piston 4 cuts off the low-load advance can be effected by means of the screw 10 in the same illustration. At pressures above this set point, only the speed-controlled advance system is in operation.

At low speed and light load, the transfer pressure acting on the low-load advance piston 4 is is insufficient to overcome the pre-load in its return spring so it remains seated. The low-load advance spindle 2, actuated by lever 7 in Fig. 6.33, is in a position such that the passages B, D and E in Fig 6.34 are all open. Consequently, the excess fuel delivery valve and the

DISTRIBUTOR TYPE INJECTION PUMPS

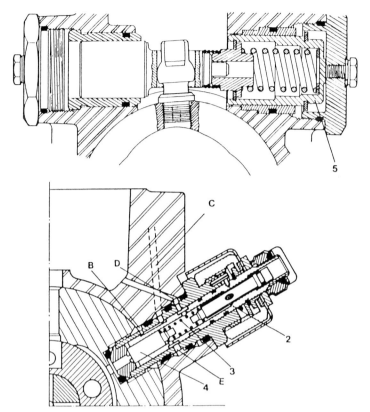

Fig. 6.34 The low load advance valve in the condition required for idling or fast idling speed 2 Low load advance shaft, as in Fig. 6.33, 3 Valve spring, 4 Valve piston, 5 Advance piston B, D and E are ducts as described in the text

chamber behind the low-load advance piston 5 are interconnected, through B and C, and subjected to the transfer pressure, through E. In this condition some of the transfer pressure is bled through D to the pump housing, so the pressure behind 5 is too little to overcome its return spring pre-load, and the injection is therefore advanced.

If the load at low speed is increased, further depression of accelerator pedal by the driver, to keep the speed constant, rotates the low load advance spindle 2 until the port E is closed. Consequently, transfer pressure builds up through B and C behind the advance piston 5 and, since its area is greater than that of the speed-controlled piston opposite, pushes it into the full load position, leaving the advance varied by only the speed controlled piston.

As the speed is increased, transfer pressure acting on the low-load advance piston 4 rises above that needed to overcome the pre-load in its return spring 3. The consequent displacement of the piston cuts off port E, isolating it from D. In this condition, regardless of the actual position of the accelerator pedal, transfer pressure can pass through only ports B and C, maintaining

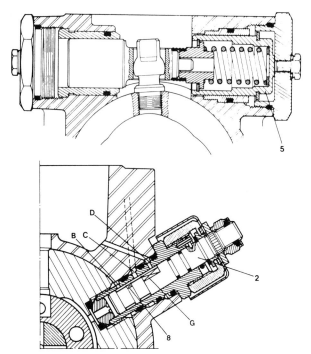

Fig. 6.35 Low load advance valve arrangement for use where speed dependent cut off is not required. Here, the fixed orifice in Fig. 6.30 is replaced by the flat 8 on the shaft, for interconnecting the ports B and C, and another flat, G, for connecting these two ports with D

the low-load advance piston 5 in the full load position. Consequently, again, only the speed-controlled advance is in operation.

The foregoing description applies if the low-load advance is to be cut-off at a predetermined speed. If it is not, a slightly different valve arrangement is used, without a speed-dependent low-load advance cut-off. The fixed orifice in Fig. 6.30 is replaced by a flat 8 on the shaft, Fig. 6.35, to interconnect the two ports B and C.

In the low-load condition, another flat, G, is in line with the discharge port D to the pump housing, to interconnect passages B and C, from the excess fuel delivery valve and to the low-load advance device respectively, with D. This allows transfer pressure to be bled into the pump housing, so the pressure behind the low-load advance piston 5 falls, to advance the injection for low-load operation.

In the full load position, when the throttle lever has rotated the valve until the flat G is isolated from the port D, transfer pressure builds up through passages B and C behind the low load advance piston 5. This pushes it into the full load position.

A boost controller can be fitted to the DPC pumps, to adjust the maximum delivery from the pumping element relatively to the boost pressure. When it is fitted, a modified fuel delivery carriage must be used, Fig. 6.36. Instead

DISTRIBUTOR TYPE INJECTION PUMPS

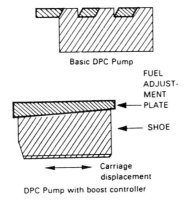

Fig. 6.36 Maximum delivery limiting devices: top, for use on the DPC pump without and, below, with a boost controller

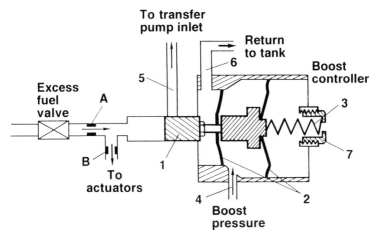

Fig. 6.37 Diagrammatic illustration of boost controller at rest or idling. The key letters are referred to in the text

of slots machined in the roller shoes, their inner surfaces are inclined to match similar inclines on the fuel adjustment plate.

Transfer pressure, opening the excess fuel valve, passes through the supply orifice A to both the piston 1, in Fig. 6.37, and through actuator circuit orifice B, where it becomes the *actuator pressure*, to the actuators which, in the naturally aspirated system, are the excess fuel delivery pistons, 4 in Fig. 6.25. As before, the actuators move the fuel delivery carriage against the excess fuel delivery spring. In this case, however, actuator pressure is the transfer pressure until a signal from the boost pressure sensor causes it to be reduced. The boost pressure unit, which contains two diaphragms of different diameters, is on the right in Fig. 6.37.

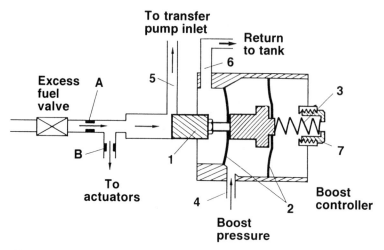

Fig. 6.38 Boost controller with the engine running at high speed

At cranking speeds, the excess fuel valve is closed, cutting off the transfer pressure from the actuators. Therefore, the excess fuel delivery spring moves the delivery carriage over to the excess fuel position, for starting the engine, and the boost controller is inoperative. As soon as the engine is running at low speed, the transfer pressure opens the excess fuel valve and therefore is transmitted two ways: through orifice A to the piston 1 and through orifice B, where it becomes actuator pressure, to the actuators. Since, at this speed, the turbocharger is not delivering any significant boost pressure, the actuator pressure moves the fuel carriage to the minimum fuel position.

As engine speed increases, the rising transfer and boost pressures act on the piston 1 in Fig. 6.38. Transfer pressure alone, however, without the contribution from the boost pressure unit, would not be enough to deflect the spring 3 against its pre-load, so the discharge port 5 is not uncovered until the boost pressure, acting on the two diaphragms 2, is high enough. Since the diaphragm adjacent to the spring is of larger area than the other, the resultant force compresses the spring and moves the piston to the right, uncovering the discharge port. As a result, the actuator pressure is reduced and the spring moves the fuel carriage assembly towards the excess fuel position until, at maximum boost, it is in the same position as for starting. A sectioned boost controller is shown in Fig. 6.39.

6.13 Electronic control of distributor pumps—Lucas EPIC system

As the demands in respect of performance, economy, driveability, general refinement, and emissions become increasingly stringent, the conventional simple and cost-effective mechanically controlled systems are becoming

DISTRIBUTOR TYPE INJECTION PUMPS

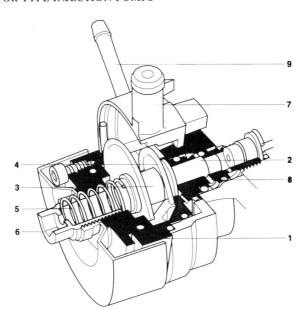

Fig. 6.39 A section through the boost controller
1 Valve body, 2 Piston, 3 Spacer, 4 Diaphragms, 5 Spring, 6 Fuel adjustment screw, 7 Boost pressure inlet, 8 Discharge to transfer pump inlet, 9 Backleak return

inadequate. Adapting them to the changing conditions renders them complex and costly. Also, the inevitable additional complexity introduces backlash, friction and more components subject to wear, and entails the addition of electronic system for limiting emissions by trimming the control.

With the development and production in large quantities of electronic controls for engine management in the 1970s, the costs of electronic components and systems fell. Moreover, because electronics are more flexible than either mechanical or hydraulic controls, it became obvious to Lucas that the extremely high degree of optimisation of fuelling essential for meeting modern requirements for diesel engines could be attained only by electronic control. Their design and development work on the EPIC (Electronic Programmed Injection Control) system was first considered in the late 1970s, and development began in the 1980s. However, it was not until 1991 that the system was first installed in a production engine.

This system had been originally intended for light duty applications. Therefore, it was based on the distributor pump, and for engines of 3, 4 and 6 cylinders up to 0.75 litres per cylinder. However, because of the trend towards direct injection for engines in this range, the pump was designed so that it could be adapted for HSDI applications running at up to 4500 rev/min. With injection pressures of about 1100 bar, it would therefore be suitable also for medium duty engines, which require higher rates of fuelling but mostly operating up to only about 3500 rev/min. For these high pressure applications, a rotor carrying four plungers is installed. With such an

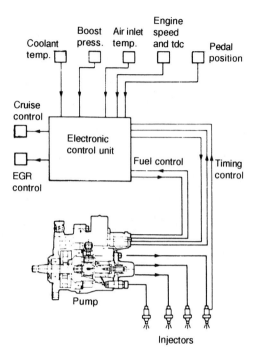

Fig. 6.40 Lucas EPIC electronic control system for indirect injection

arrangement, it is possible to shape the rate of injection by causing one pair of plungers to strike their cams earlier than the other two while arranging for all four to retract together.

Direct injection versions do not have the injector needle sensor indicated in Fig. 6.40. Instead, a resistance sensor monitoring cam position is installed as an option. Also, a Hall effect sensor detecting the passage of an assymetric vane is attached to the driveshaft. This, by taking into account the precise angular position in which the pump is bolted to the shaft, enables the ECU to correct injection timing in relation to manufacturing and assembly tolerances. This sensor also provides a back-up signal in the unlikely event of failure of the engine speed sensor.

The injection pump, Fig. 6.41, is similar to that of the DPC, but with some major changes to render it suitable for electronic control. Notable among these changes are the siting of the transfer pump on the drive shaft, in the position formerly occupied by the mechanical governor. This leaves the other end of the pump rotor shaft clear for the installation of a sensor for monitoring the axial position of the rotor. The axially movement of the rotor, which is effected by an electro-hydraulic actuator, controls the fuelling by sliding the the cam roller follower shoes up and down an inclined surface, Fig. 6.42.

Because the rotor moves axially to control the plunger stroke over the whole delivery range, the axially sliding carriage of the DPC is not needed.

DISTRIBUTOR TYPE INJECTION PUMPS

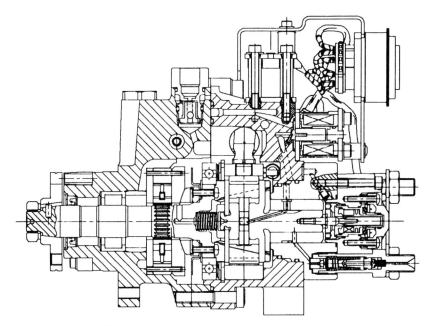

Fig. 6.41 Lucas EPIC electronically controlled distributor type injection pump

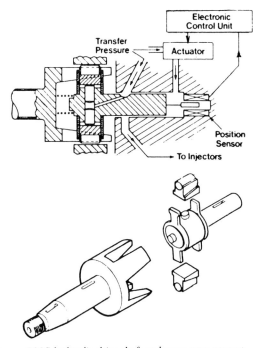

Fig. 6.42 Top, Lucas EPIC hydraulic drive shaft and rotor arrangement

As before, transfer pump pressure forces the plungers apart, to fill the chambers in the rotor, and cams on a cam ring return them, to deliver the fuel at the high pressure needed for injection. Although the timing advance unit is similar to that of the DPC it has been modified in detail for electro-hydraulic actuation.

A single short hole drilled in the distributor shaft serves for both the filling of and delivery from the injection pump Fig. 6.41 and 6.42. By virtue of the consequently very small volume of fuel between the pumping plungers and the injectors, variations in delivery owing to compressibility of the fuel are minimised.

These changes have simplified and therefore reduced the cost of the pump, but of course do not totally offset the increased cost arising from the addition of the electric and electronic components. Even so, the weight of the unit has been reduced by between 20% and 50%, depending on which of the other pumps it is compared with, the electronically controlled pump is shorter by 28%, and the number of components has been reduced by 25%.

6.14 Application and benefits of electronic control

The EPIC closed loop, digital electronic fuelling and timing control system is illustrated in Fig. 6.40. A diagram showing the hydraulic system is shown in Fig. 6.43.

Comprising surface mounted and standard components on printed circuit boards, the electronic control unit (ECU) can be easily configured to accommodate features such as EGR, turbocharger, traction control, and cruise control and other driver aids. It embraces a single chip microprocessor incorporating the CPU, ROM, RAM, A/D converter, timers and digital I/O circuitry. All input signals and closed loop controls are monitored continually, for correct range and validity, by the microprocessor. The information thus logged can be utilised for diagnostic purposes or, for example if some sensors have failed, to switch over to recovery strategies. The digital signals are used as inputs to a number of maps, such as that in Fig. 6.44, and tables stored in the memory.

To ensure accuracy under idling and transient conditions, speed and crankshaft angle signals are obtained from a variable reluctance transducer, sensing grooves or pegs on the engine flywheel, and applied in association with signals from a variable reluctance sensor in one of the injectors. The accelerator position sensor is a plastics film potentiometer combined with a zero pedal angle switch, which is interrogated under zero demand conditions to check the analogue reading. Thermistors are used for measurement of temperatures, and a piezo-resistive sensor measures boost pressure. For the maintenance of constant mass of fuel injected, the thermistor for measuring fuel temperature is sited in the injection pump cam position sensor.

The curve of delivery against rotor position is linear for IDI but mapped for DI engines. To ensure that the rotor control is insensitive to variations in fuel viscosity or between actuators, an adaptive strategy is employed.

DISTRIBUTOR TYPE INJECTION PUMPS

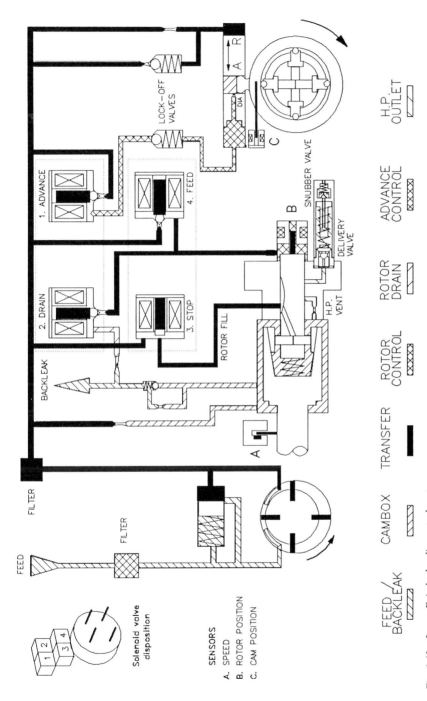

Fig. 6.43 *Lucas Epic hydraulic control system*

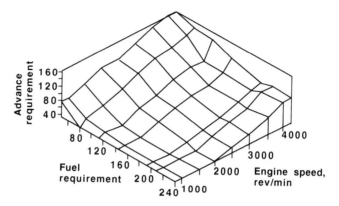

Fig. 6.44 Typical map of engine fuel and injection advance requirements against speed

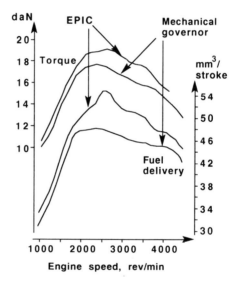

Fig. 6.45 By taking advantage of the potential for matching the fuel delivery more closely to engine requirements with EPIC than with a mechanically governed system, it is possible significantly to improve the torque characteristics

Consequently, the degree of accuracy of control of the rotor, required for good driveability and inter-cylinder balance, is achieved. Furthermore, by virtue of the flexibility of mapping for fuel delivery curve shape and advance, the system can be matched very closely to the engine requirements, Fig. 6.45. Provision is made in the ECU for accurate control over boost pressure by means of either a wastegate or variable geometry vanes in the turbocharger.

With improvements in engine mountings, it is becoming possible to reduce diesel engine idling speeds. Moreover, by virtue of the EPIC electronic control system, with its isochronous idle governing, feedback and close control over

DISTRIBUTOR TYPE INJECTION PUMPS

the delivery strokes, together with consistency of line-to-line delivery, the attainment of low idling speeds and maintenance of stable speed under both steady state and varying loads is practicable.

Repetitive factors such as variations in nozzle opening pressure, and compression and friction from cylinder-to-cylinder, cause periodic changes in rev/min at a frequency of half engine speed under idling conditions. However, by virtue of the adaptive control software, it is possible to measure the mean speed relative to the combustion event in each cylinder, and to vary cyclically the quantity of fuel injected, to keep the engine running constantly at the reference speed.

With electronic control, not only can the speed be kept constant, but also noise, HC emissions and smoke minimised, and all regardless of changes in temperature following a start from cold. Additionally it is possible, at all speeds and temperatures, to reduce torsional oscillations due to variations in instantaneous engine speed, by phasing the fuelling in opposition to the speed changes. Such oscillations are tending to become particularly troublesome as the ratios of engine torque to vehicle weight increase. Closed loop control of injection advance, based on the use of an injector needle lift sensor, reduces the scatter, and hence variations of emissions between individual cars, due to the fitting of different combinations of engine and pump.

Should it prove to be too difficult or costly to install a needle lift sensor, the alternative of a combination of the crankshaft and pump cam position sensors can be used, Fig. 6.46. As previously mentioned, EGR control can be exercised by the system. However, by virtue of the accuracy of instantaneous fuelling obtainable, it may even be possible to avoid the installation of EGR equipment on engines of less than 2 litres capacity.

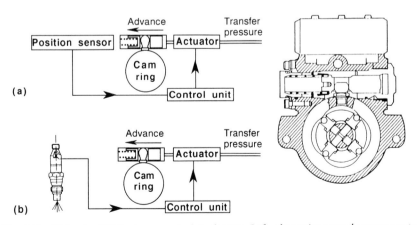

Fig. 6.46 Right, the EPIC advance control mechanism. Left, alternative control arrangements: above with crankshaft and pump position sensors; below, with needle lift sensor

6.15 The Stanadyne distributor pumps

In 1947, Vernon Roosa brought to the Hartford Division of Stanadyne his new concept of substituting a single pumping rotor, in which were two opposed plungers serving all cylinders, for the then universal in-line injection pump with an individual plunger for each cylinder. Another new feature of his pump was inlet instead of spill metering, which made the pump almost self governing, so that it required only a simple low cost governor. It was the smallest and simplest injection pump ever, and therefore could be produced at a lower cost than the in-line pumps.

In 1952, after 5 years of development, the Roosa Master, Model A pump, with mechanical governing, was put into production for the Hercules Motors Corporation, for fitting to the Oliver Cletrac tractors. In 1953, it was supplied to Continental Motors and the following year to the Buda Engine Co, later to become Allis Chalmers. During the period 1955 to 1958, the Model B and D pumps were introduced, the former with a sand cast and the latter with a diecast housing. In 1958, the Model DB was announced as the successor to the A and D versions. A heavier duty pump, the DM, was introduced in 1972. It had a heavier section rotor, four plungers and a new hydraulic head configuration. The DB2, first produced in 1977, is the second generation DB pump. The pump is illustrated in Fig. 6.47, and a schematic diagram of the whole system in Fig. 6.48.

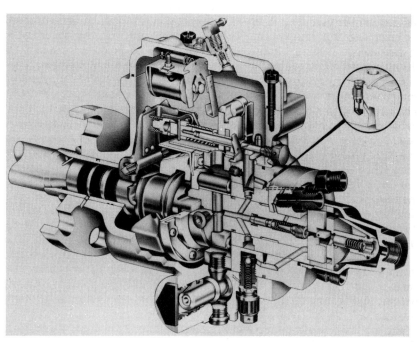

Fig. 6.47 The Stanadyne DB2 pump with a solenoid-actuated mechanism in the top cover for key start and stoop operation

DISTRIBUTOR TYPE INJECTION PUMPS

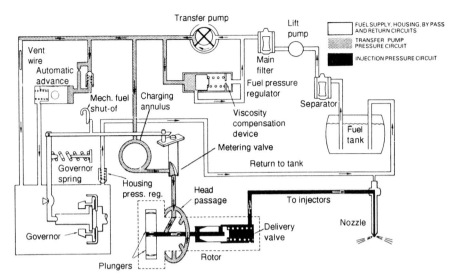

Fig. 6.48 Schematic diagram of the Stanadyne DB2 system

In general the DB2 pump is similar to the Lucas DP series originally produced under a licence from Stanadyne, as explained in Section 6.1. The differences, however, are of considerable interest. As can be seen from Fig. 6.48, the centrifugal governor differs only in detail layout. Either all-speed or two-speed governing can be arranged, on principles similar to those of the Lucas DPS and DPC units respectively described in Sections 6.8 and 6.11.

An interesting difference, however, is the provision for fuel temperature compensation on the two-speed governed version, illustrated in Fig. 6.49. The rate of change in fuel flow with temperature amounts to about 0.9% by weight and 1.8% by volume per 10 deg C (0.5% and 1% respectively per 10 deg F). When the pump is mounted between the banks of cylinders in a V-engine, fuel temperatures can become particularly high, and therefore compensation especially desirable. As temperature increases, idling speed decreases. However, setting the idling speed high enough to prevent stalling may be impracticable, since it can cause automatic transmission creep. For this reason, a bimetal strip is mounted in series with the idling spring and biased in a manner such that, as the temperature of the fuel in the pump rises, it increases the open area of the fuel metering valve.

During rapid acceleration at low temperatures, the shearing of the fuel film in the head generates a great deal of heat. Because the distributor shaft is smaller than the sleeve in which it rotates in the hydraulic head, it may become even hotter and expand more rapidly. This can cause the clearance between rotor and sleeve to close and cause seizure. The problem arises where the temperature is highest, which is approximately mid way between the ends of the bearing surfaces. Stanadyne have found that machining a peripheral groove around the affected area of shaft can be the solution.

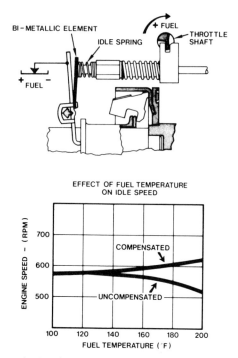

Fig. 6.49 Above, the mechanism for temperature compensation of engine idling speed. Below, the effect of temperature compensation

In other respects, the rotor, Fig. 6.50, is much the same as that of the Lucas DPC pump, Section 6.11, but the fuel delivery ducting is different, the maximum fuel adjustment plate is described by Stanadyne as a leaf spring, and there are no slots in its ends. Locking of the adjuster screw that is tightened on to the spring plate is obtained by friction, supplemented by a special coating material on its threads.

Stanadyne did some development work on an interesting variant of the spring plate arrangement, Fig. 6.51, but did not put it into production. It comprised two leaf-spring, fuel adjustment plates, each pivoting about its adjustment screw. The outer end of each limited the travel of the plungers, while the inner ends loaded a piston sliding in a radial hole mid-way between them. Fuel transfer pressure forced the pistons outwards against the inward load applied by coil springs.

When the engine is idling and the transfer pressure therefore low, each piston is in its innermost position. The inner ends of the fuel adjustment plates can follow them, so they pivot about their screws to allow the plungers their maximum outward travel. As the engine speed increases, the rising fuel transfer pressure forces the piston outwards, pivoting the plates in the opposite direction about their screws, and therefore pulling their outer ends inwards, to reduce the maximum fuelling.

DISTRIBUTOR TYPE INJECTION PUMPS

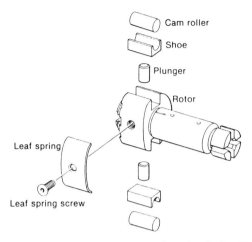

Fig. 6.50 The DB2 rotor is similar to the Lucas equivalent already described in Section 6.11

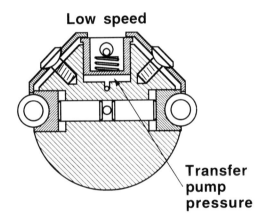

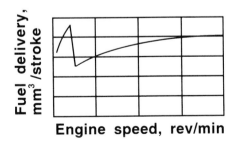

Fig. 6.51 This experimental version of the spring plate arrangement was not put into production

162 DISTRIBUTOR TYPE INJECTION PUMPS

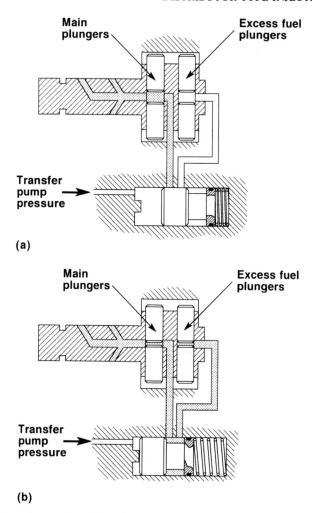

Fig. 6.52 Stanadyne developed four-plunger variant of the pump, for operation in very cold climates, but it never went into production. At (a), only two plungers are effective, for normal operation while, at (b) all four are working, for operation with excess fuel

For operation in very cold climates, where starting may be exceptionally difficult, an injection pump rotor with tandem plungers has been developed for use as an excess fuel device, Fig. 6.52, but again this was not put into production. A spool type control valve is situated in the rotor head, opposite the maximum fuel leaf-spring. When the transfer pressure is low, a return spring opens the control valve so that both pairs of plungers are in operation. As the engine fires and picks up speed, the rising transfer pressure closes the spool valve leaving only one pair of plungers in operation. Since the extra

DISTRIBUTOR TYPE INJECTION PUMPS

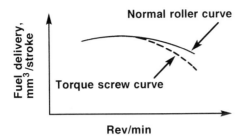

Fig. 6.53 A manually adjustable restrictor in the flow to the metering valve produces this torque back-up effect

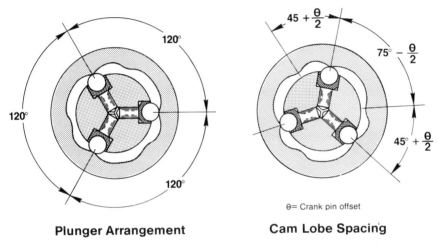

Fig. 6.54 This illustrates another development that was never put into production. Three opposed pairs of plungers were set at 120 deg, to cater for 90 deg V6 engines

fuel supplied may be between 80% and 140% above normal, the pressure at which it is cut off is well below that generated at low idling speed.

To improve torque back-up (by increasing torque with decreasing speed), a torque screw is employed. This is in effect simply a restrictor limiting the flow through the metering valve. This restrictor is calibrated to set both the rated speed and the limit on maximum power. As the speed and therefore transfer pressure falls, the effect of the increase in time available for injection is greater than that of restriction of flow. From the point at which maximum torque is developed, which is determined by the maximum plunger stroke, up to the rated speed, the torque curve is almost a straight line, Fig. 6.53.

Another interesting development that has not gone into production is a plunger and cam ring layout, for uneven firing 90 deg V-six engines, Fig. 6.54. Three pairs of plungers are set at 120 deg but the cam lobe setting depends on the crank pin angle. The distributor ports are arranged to suit the uneven firing intervals.

6.16 Stanadyne fuel delivery arrangements

As indicated in the previous Section, the arrangement of the fuel inlet ducting to the pump plungers is similar to that of the Lucas DPC pump. However, the high pressure output from the plungers passes axially through the distributor rotor to the end, which is counterbored to receive a cylindrical delivery valve. During rotation, the outer end of a duct drilled from the periphery of the spring chamber behind the valve, Fig. 6.55, aligns in turn with each of the ports leading to the injectors.

The delivery valve has a snubbing action in that, as it retracts towards its seat, the end opposite to the seat withdraws into the valve bore, closing the radial ports in its periphery. Further motion towards its seat, that is through the distance h, causes a negative pressure wave to be transmitted to the injector, lowering the line pressure and thus ensuring that the injector needle seats rapidly and firmly without secondary injection or dribbling.

In some applications, especially if the retraction volume exceeds 30 mm^3, the action of the valve on closing can cause fuel column separation, producing vapour-filled cavities, just downstream of the delivery valve. To overcome this problem, a pressurising valve could have been incorporated in the delivery pipe connector, though Stanadyne decided not to go ahead with this idea. The action of this valve was, in principle, similar to that of the delivery valve, but its displacement was significantly larger. When the two valves were returned by their springs towards their seats, the delivery valve seated first owing to its smaller displacement. The pressurising valve then applied to the delivery valve spring chamber a pressure that was equal to the pressurising

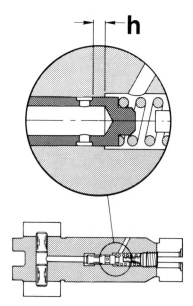

Fig. 6.55   This delivery valve is housed in a counterbore in the end of the distributor rotor

DISTRIBUTOR TYPE INJECTION PUMPS

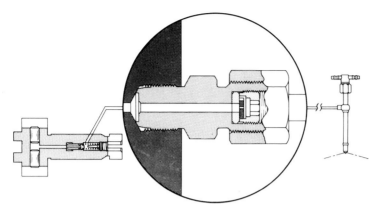

Fig. 6.56 On the DB2 pump, snubber valves are installed in the pipe connections to the cylinders. The small axial hole through the valve damps reflected waves to reduce the potential for cavitation erosion

valve spring pre-load divided by its area. This ensured that vapour cavities could not form in the delivery valve spring chamber.

An alternative is the fitting of snubber valves in the pipe connectors, Fig. 6.56. This is simply a plate type non-return valve with a small orifice in its centre. Delivery of fuel to the injector opens the valve which, when delivery ceases, is closed again by the falling pressure, but the negative pressure wave is damped as it passes through the snubber valve orifice, thus reducing the cavitation erosion potential. This orifice also damps reflected pressure waves, so secondary injection can be avoided without having to have a large retraction volume in the delivery valve. In extreme circumstances, owing to difficulties at cranking and idling speeds, it may be necessary to fit a snubber valve downstream within the pressurising valve housing.

The injection timing advance mechanism, with a piston acted upon at one end by transfer pressure and at the other by a return spring, rotates the cam ring. In one version, the piston is fitted with an iron sealing ring. A check-valve locks the mechanism hydraulically against reverse rotation by the shock of the rollers striking the cams. In this pump, a screw is provided for adjusting the pre-load in the return spring, and an additional advance mechanism is incorporated to adjust automatically the timing in relation to speed throughout the range. The maximum advance available is 10 deg of pump shaft rotation for eight cylinder engines and 12 deg for the others. However, if the fuel delivery is less than 30 mm^3 per stroke, this can be increased by 1 or 2 deg. A servo controlled advance is also available.

The advance mechanism compensates for two effects: one is the injection delay arising owing to the time needed for the pressure wave to travel from pump plungers to injectors, and the second is the ignition delay period. The formula used by Stanadyne to calculate this delay is:

$$\text{Cam advance} = \frac{L(N_2 - N_1)}{16\,800}$$

where L = length of line in inches
   N_2 = rated speed
   N_1 = minimum full load speed.

The assumptions are that there are no vapour cavities in the line and that the wave speed is 4200 ft/sec (1280.16 m/s). If metric units are employed throughout, the figure of 16800 becomes 130.064.

6.17 Stanadyne DS pump

Mechanically, this electronically controlled pump introduced in 1993 is similar to the DB2 unit just described but of course without the governor. For high speed 4-cylinder applications, it is capable of delivering 75 mm^3 of fuel per stroke, at a maximum presssure of 1200 bar at the injectors. The recommended maximum size of engine cylinder is 0.5 litres.

In common with the other pumps of this type, the quantity of fuel delivered to the plungers for injection into the cylinders is determined by an electronically controlled, solenoid-actuated spill valve of the poppet type. Stanadyne have also incorporated a stepper motor to actuate the cam ring advance mechanism. Consequently, both the quantity of fuel injected and the timing of the start of injection are accurately regulated, in relation to load, speed, and other engine parameters to keep emissions of HC, CO, NO$_x$ and smoke to a minimum under all conditions of operation.

The layout of the pump can be seen from Fig. 6.57. A unique feature of the spill valve is that it is housed coaxially in a counterbore in the end of the distributor rotor. This arrangement offers a major advantage in that the volume of fuel subjected to the injection pressures is very small, so higher pressures can be adopted without risk of compressibility effects significantly affecting injection characteristics, which therefore accurately reflect the cam geometry.

Further enhancement of the accuracy of control is assured by the housing of the cam rollers and tappets (actuating four plungers) in a large diameter, zero backlash, drive shaft. As can be seen from the illustration, this leaves the distributor rotor isolated from the driving loads, thus ensuring that it is unaffected by torsional deflections which otherwise, might lead to inaccuracies in the timing. Provision is made for accommodating a high capacity belt drive if required.

The control system is illustrated diagrammatically in Fig. 6.58. Data supplied to the electronic control module (ECM) by the engine-mounted sensors is continuously updated, as also are those related to pump speed and angular pulse train. These signals are processed by custom alogorithms, so that the appropriate command signals can be sent to a pump mounted solenoid driver (PMD) and cam ring advance stepper motor.

By virtue of the use of the single, high speed solenoid, the benefits of ease, flexibility and accuracy of signal processing associated with digital control have been obtainable. Fuel metering and timing are regulated as a function of input data supplied to the ECM. This module controls the PMD, which

DISTRIBUTOR TYPE INJECTION PUMPS

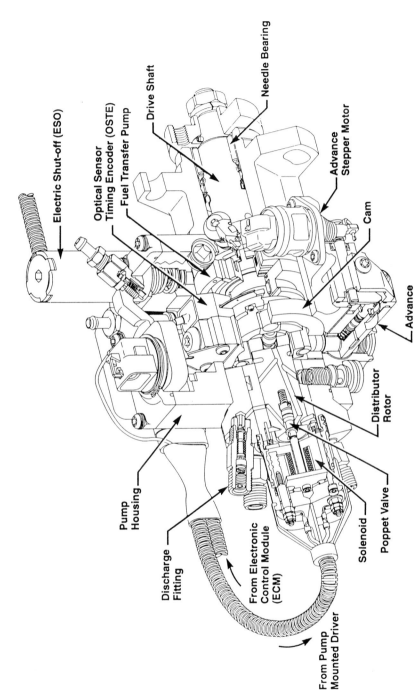

Fig. 6.57 *The Stanadyne DS electronically controlled pump was introduced late in 1993*

DISTRIBUTOR TYPE INJECTION PUMPS

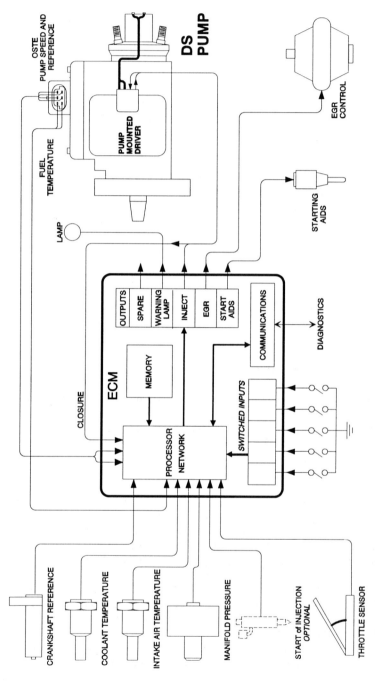

Fig. 6.58 Schematic diagram of the DS electronic control system

provides it with a constant current and the injection command signals. Closure of the poppet type spill valve is detected by the PMD, and the event signalled back to the ECM. Fuel quantity required for injection, varied if necessary between 0 and 100%, and its timing are updated for each successive injection, so engine response to changes in load and speed is virtually instantaneous.

A significant feature of the control strategy is that it is angle instead of time based. This gives superior performance because both metering and timing requirements are a function of crankshaft angle. Stanadyne engineers have found this feature to be particularly advantageous in transient conditions of operation.

The function of a high resolution angular clock is provided by an encoder mounted on the pump drive shaft. Its output is enhanced by a phase lock loop (PLL) circuit in the ECM, giving a resolution of 0.04 deg. The fuel metering and timing events are controlled, on the basis of the signals received from the angular clock, by a series of digital counters in the ECM.

A useful minor feature of the equipment is the situation of the fuel inlet at the top, where it is readily accessible even on V-engines. The principal advantages claimed by Stanadyne for their new pump, however, are the following: heavy duty drive components, flexibility of fuel metering and timing control on a shot-to-shot basis, and therefore good performance in transient conditions, flexibility of governing and idle speed control, flexibility of control for cold running.

6.18 The Bosch VE distributor type injection pump

As regards fuel distribution principle, the Bosch VE injection pump, Fig. 6.59, is similar to the others already described, but is totally different in mechanical detail. Keyed on to the input shaft, immediately inside the cover, is the eccentric vane type transfer pump rotor assembly. It runs in an eccentric sleeve in the housing, into the top of which is screwed the pressure control valve.

As the pressure control valve is lifted, against the pre-load in a return spring, it progressively uncovers a port through which the delivery in excess of requirements passes into the pump casing. At the top of the casing, a restrictor in an outlet to the tank ensures that the transfer pressure is proportional to speed. If the transfer pressure rises above a pre-set maximum, the valve lifts further, uncovering a port through which the delivery excess to requirements is returned to the inlet.

Next to the transfer pump is a gear meshing with a pinion on the parallel governor shaft above. Immediately behind the gear, the input shaft terminates and two lugs on its end register in slots in a four-arm spider, or yoke. Two more lugs, on the adjacent flanged hub of the cam ring carrier flange, project into the other two slots in the yoke, to transfer the drive to a combined pump plunger and distributor rotor assembly. Around the spider, between the face cams and the governor drive gear, is a carrier ring for the rollers

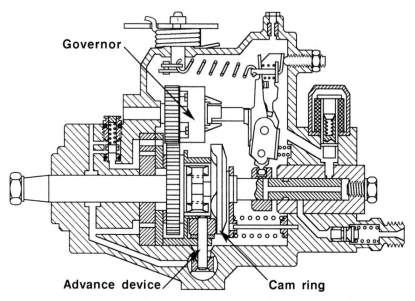

Fig. 6.59 The Bosch type VE distributor type injection pump has a single axially reciprocating plunger. The governor, by virtue of its being geared up, is commendably compact. Installed in the top of the front end of the cambox is the pressure control valve and, at the rear, a solenoid actuated fuel shut-off valve. Mechanical actuation of the shut-off valve is an alternative

that serve as the cam followers. This ring is fixed to the housing. The face cam ring is mounted on the same side of the flange as the lugs that drive it.

The other side of the hub of the cam ring carrier is counterbored to receive a flange on the adjacent end of the combined pump plunger and distributor shaft. This flange, spigoted into the counterbore, is driven by a peg pressed into a hole from the other side of the cam carrier flange, to register in a radial slot in the flange spigot.

The distributor shaft is advanced and retracted by the face cams, to serve also as the pumping plunger. Injection timing is varied, in relation to speed, in a manner similar to that of the previously described rotary injection pumps, except in that it is the roller carrier, instead of the cam ring, that is rotated by the spring-loaded piston. Timing begins to revert from retard to normal when the engine speed rises to about 300 rev/min. The maximum angle of rotation of the roller carrier can be as much as 12 deg, which of course is 24 deg crankshaft rotation.

From Fig. 6.60, it can be seen that the mechanism for advancing and retarding injection, though rotating the roller carrier ring, is broadly similar to those of the other pumps described. With the pump running at half engine speed on a four stroke engine, there are as many cams and rollers as cylinders.

So that the cam ring will ride constantly on the rollers, the cam ring assembly is loaded axially by a pair of coil springs, one each side of the rotor. These springs are compressed between the distributor head and the

DISTRIBUTOR TYPE INJECTION PUMPS

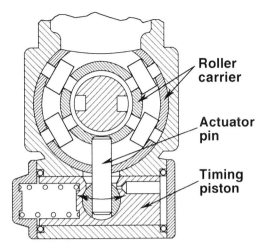

Fig. 6.60 This Bosch mechanism for advancing and retarding injection is similar to those already described, but it rotates the roller carrier ring instead of a cam ring

two arms of a yoke, the centre of which straddles the rotor and seats on the spigot flange.

The flange on the end of the rotor is spigoted in and secured to the adjacent end of the hub of the cam ring carrier. Consequently, each time a diametrically opposite pair of the face cams rides over the rollers, the distributor rotor is moved axially to the rear, to actuate the pumping plunger which, at the same time, is rotating.

How the fuel is metered and distributed can be seen from Fig. 6.61. Between the yoke and the distributor head is a control sleeve which, under the control of the governor mechanism, is slid axially along the shaft to vary the spill point, at which the plunger ceases to deliver to the injectors. As the peaks of the cams leave the rollers, and the rotating plunger therefore begins to retract, fuel under transfer pressure goes through the inlet into the metering slit, to enter the high pressure chamber. When the next pair of cams rides over the rollers, the metering slit has closed and the distributor slit has opened to the delivery port, so fuel is forced from the high pressure chamber into the axial hole and then radially outwards into the delivery port to the appropriate injector. Injection is terminated when the control sleeve uncovers the spill port, the position of which is determined by the governor.

The fuel is delivered through what Bosch term a *pressure valve*, Fig. 6.62 screwed into the outlet port to the high pressure line to the injector. Similar to the delivery valves already described in connection with the other pumps, it is a fluted valve with an unloading collar a short distance beneath its conical seat. Incorporated in it however, but not shown in the illustration, is a snubber valve with a restrictor orifice in its centre. This is a plate valve normally held on its seat by a return spring, but is lifted by the delivery pressure in the delivery line. When it is seated again, following the opening of the spill port, the restrictor orifice damps the pressure waves generated

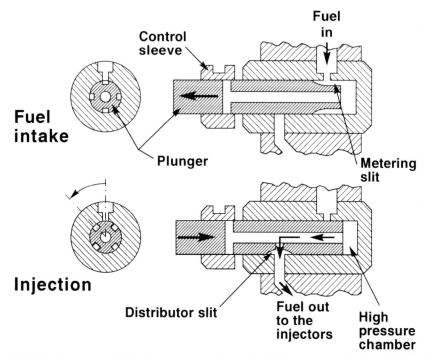

Fig. 6.61 Control over fuelling by the governor in the Bosch VE type pumps is effected by axial movement of the control sleeve to vary the spill point

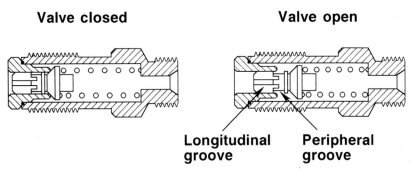

Fig. 6.62 The Bosch pressure valve (delivery valve) incorporates an unloading collar and, not shown in this illustration, a snubber valve in the form of a spring-loaded plate with a restriction orifice in its centre

by the sudden seating of the injector needle. The aim of course is at preventing post-injection due to reflection of these waves back to the needle.

Incidentally, for very many years, Bosch have been producing distributor type pumps with only a single pumping element coaxial with the shaft. At the time of going to press however, they have announced that they are to produce some new types in 4 variants, one with radially disposed plungers for DI engines having 3, 4 or 6 cylinders delivers at pressures up to 1350 bar at the nozzles. This is termed the VR...M type. The others, the VE...M Series, retains the axial plunger, delivers at up to 1000 bar the injectors and is intended for both IDI and DI engines. Benefits claimed include high accuracy, wide range of adjustment of start of injection, variable delivery rate and cylinder specific correction of quantity of fuel injected. With two-stage injection, noise reduction too is claimed.

6.19 Governing the Bosch VE type pump

As mentioned previously, the governor is gear driven on a separate shaft parallel to and above the main shaft, to operate at about 1 1/2 times rotor speed. The all-speed version is illustrated in Fig. 6.63. When the engine is stopped, the starting lever is held in its starting position by the leaf spring. This holds the control collar in the pumping and distributor plunger in its rich position, in which the spill port is uncovered only when the plunger stroke is at its maximum.

When the engine starts, the governor weights slide the sleeve axially, bending the leaf of the relatively weak starting spring until the starting lever comes up against its stop on the what Bosch term *the tensioning lever*. This moves the control collar back to open the spill port earlier, so the fuelling rate drops back to that for idling. In this condition, the idling speed is kept constant by the idling spring as follows.

The governor spring is connected at one end to the control lever and, at the other, to a shouldered pin passed through a hole in the main vertical lever which, as mentioned previously, is termed the *tensioning lever*. Interposed between the head of this pin and the tensioning lever is the idling spring. Consequently, when the control lever is moved to increase the speed, it compresses the idling spring progressively, until the shoulder on the pin comes up against the tensioning lever. Up to the point at which the idling spring is fully compressed, however, the tensioning lever, acting through the starting lever, will move the control collar to vary the spill point to maintain a constant idling speed for any given pedal angle, regardless of changes in load.

When the idling spring is fully compressed, the engine still operates at the speed dictated by the angle to which the driver sets the accelerator pedal, but the governing is then done by the main governor spring. If the load decreases, and the speed therefore tends to rise, the governor weights move the control sleeve to increase the tension in the governor spring, and thus move the levers and control collar over to reduce the effective stroke of the plunger, and vice versa if it increases. During engine braking, any increase in speed causes the governor weights to fly further out to reduce the fuelling to the level that exactly balances the load on the engine.

For two-speed governing, Fig. 6.64, the control layout is almost identical, but the governor spring becomes a compression spring located inside a cylindrical housing. Also, an additional coil spring is seated around the stop

DISTRIBUTOR TYPE INJECTION PUMPS

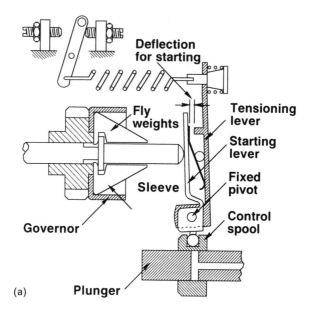

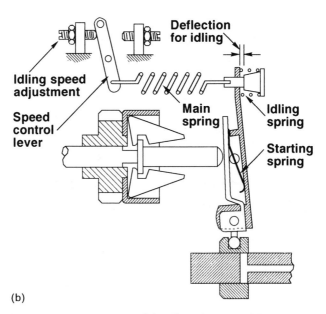

Fig. 6.63 Diagrammatic representation of the all-speed version of the Bosch governor for the VE Series pumps: (a) during starting; (b) for idling

DISTRIBUTOR TYPE INJECTION PUMPS

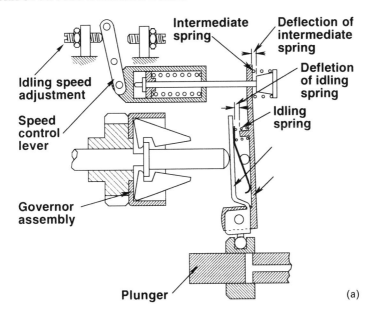

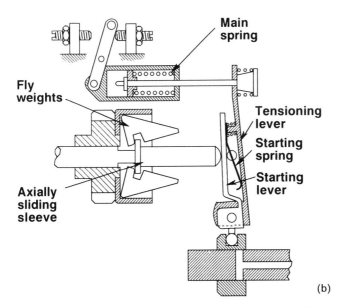

Fig. 6.64 Arrangement of the Bosch two-speed (idling and maximum speed) version of the governor: (a) idling: (b) at full load

for the starting lever. This becomes the idling spring, while that which formerly served as the idling spring is now termed the *intermediate spring*. The starting spring remains the same. Consequently, for starting, it presses the starting lever against the end of the governor sleeve, moving the control collar to the maximum fuelling position.

When the engine starts, constant idling speed is maintained automatically by balancing the force exerted on the control sleeve by the governor weights and the reaction of the idling spring to it. When movement of the accelerator pedal fully compresses the idling spring, the idling lever comes against its stop on the tensioning lever. At this point, the intermediate spring comes into effect, still maintaining a constant speed for any given pedal angle. This provides an extension to the idling speed range, a large droop, and a smooth transition to the next stage.

Because the pre-load in the governor spring is too high to be compressed at this stage, control over fuelling under light load conditions is effected directly by movement of the accelerator pedal right up to rated speed. If, at any point, the engine speed begins to rise above its rated maximum, the governor weights, under the influence of centrifugal force, slide the sleeve along the plunger to compress the main (or governor) spring, thus moving the control spool to reduce the rate of fuelling and prevent overspeeding. If, in the meantime, the load increases but the pedal is held at a constant angle, the reduction in speed will cause the governor weights to move inwards, holding the speed constant. When the accelerator pedal reaches its maximum travel, any further increase in load will cause the speed to fall.

6.20 Optional extras for the Bosch VE type pump

The wide range of optional extra features provided by Bosch for their VE type pump includes several alternative measures for the control over torque. As mentioned previously, torque control entails modifying the fuel delivery to compensate for the changes in engine fuelling requirements relative to speed, Fig. 6.65. Because the spill port opens increasingly rapidly with speed,

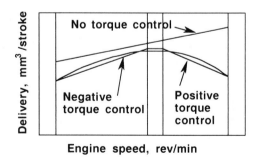

Fig. 6.65 Diagram illustrating how torque control can be utilised to influence pump delivery. The straight line at the top represents excess fuel delivery without torque control, and the curve the engine fuel requirement

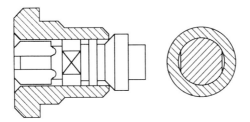

Fig. 6.66 Positive torque control can be effected most simply by means of this valve. Fuel is delivered through the clearance between the flats and the bore (section on right) so that, as the speed rises, the restriction offered by this limited clearance increasingly hinders the flow

the viscous drag on the fluid passing through it progressively decreases. Consequently, the output per stroke of the pump rises slightly more rapidly than the speed of rotation. Other factors, such a piston ring blow-by, injection timing, and rate and quality of mixture preparation affect the engine fuelling requirement, the curve of which therefore tends at first to rise more rapidly than the rate of pump delivery and then to fall off below it again. The outcome is that if the rate of fuelling is adjusted for smoke-limited maximum power output, or best torque at high speeds, it may be running too rich at low speeds and vice versa.

Negative torque control entails reducing the rate of delivery as speed is reduced. It is normally applied to engines that tend to smoke at the lower end of the speed range. Positive torque control, on the other hand, calls for reducing the rate of delivery as speed is increased.

Positive torque control is effected most simply by grinding flats on an additional collar in the delivery valve. This collar is interposed between the unloading collar and the fluted stem of the valve, Fig. 6.66, so that the fuel is delivered through the clearance between the flats and the bore of the port. At low speeds, there is plenty of time for all the fuel to pass through the flats but, as the speed rises and the time available for injection therefore becomes less, the restriction that they impose on the flow becomes increasingly significant. Consequently, the valve lifts further and the volume beneath it increases, and thus retains some of the fuel instead of delivering all of it.

A more costly, but in some respects better, way of exercising positive control over torque is to incorporate a *torque control lever* and *torque control spring*, Fig. 6.67, in the governor mechanism, Fig. 6.63. The pendant torque control lever pivots on a pin between lugs on the upper end of the starting lever. Abutting against its lower end is the head of a spring-loaded *torque control pin* but its movement to the right, as viewed in the illustration, is restricted by a lug on the tensioning lever.

The head of the torque control pin is shouldered, its smaller diameter portion sliding in a hole in the starting lever. At the opposite end, the shank of the torque control pin is passed through a hole in another lug on the starting lever, and upset to form a stop. Pre-compressed between the lug

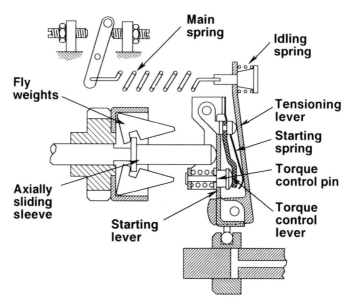

Fig. 6.67 Diagram showing a Bosch method of torque control, by incorporating a torque control lever in the governor mechanism

and the head of the pin is the torque control spring. When this spring deflects, the shank of the pin slides in the hole in the lug.

With the engine running, and the leaf type starting spring therefore collapsed, increasing speed causes the governor sleeve to push the starting lever to the right. This moves the upper pivot, common to both the starting and torque control levers, also to the right. As a result, the torque control lever pivots about the contact point between it and the lug on the tensioning lever, so its other end compresses the torque control spring, causing the small diameter portion of the head of the pin to slide to the left through the hole in the lug. Ultimately, this movement is stopped by the shoulder on the head, which is when positive torque control ceases.

Throughout the positive torque control process, the increasing force at the contact point between the torque control lever and lug on the tensioning lever progressively swings the latter lever about the lower pivot and thus slides the control collar to the left, to reduce the fuelling. The point at which torque control begins is determined by the preload in the torque control spring.

For exercising negative torque control, a slightly different torque control lever layout is needed, Fig. 6.68. This is not only because the torque control spring comes into operation later in the speed range, so it has to be placed further back from the governor spring tensioning lever, but also because the movement of the governor sleeve with increasing speed has to cause the control collar to slide in the direction opposite to that for positive torque control.

DISTRIBUTOR TYPE INJECTION PUMPS

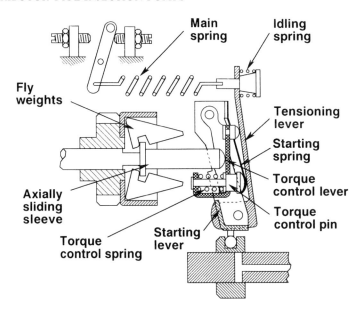

Fig. 6.68 For exercising negative torque control, a torque control lever entirely different from that in Fig. 6.67 is needed

From the illustration it can be seen that first the starting spring collapses, at which point the torque control lever comes up against the lug on the tensioning lever and the torque control pin against the tensioning lever itself. As the speed increases, the torque control lever applies an increasing load on the torque control spring. When this load exceeds the pre-load, the spring is deflected, and negative torque control begins to come into effect. The pin begins to slide axially, as before, and at the same time, the upper pivot which, again, is common to both the starting and torque control levers, swings about the lower pivot, and thus slides the control collar in the direction that increases fuelling. When further compression of the torque control spring is stopped because the shoulder on its head comes up against the torque control lever, negative torque control ceases.

In principle, the boost pressure and altitude compensation modules are similar except in that the boost pressure control is actuated by a diaphragm while altitude correction is effected by an aneroid. Consequently, it is necessary here to describe only the boost pressure module, Fig. 6.69.

Boost pressure is applied to the top of the diaphragm, which has a return spring beneath it. Secured to the centre of the diaphragm is a vertical pin the upward movement of which is limited by a screw stop. Near the bottom end of the pin, a section of its length is ground to a conical profile termed the *control cone*. As the pin is moved up and down by the diaphragm, the inner end of another pin, termed the guide pin and carried in a hole drilled radially in the guide for the vertical pin, rides up and down the conical

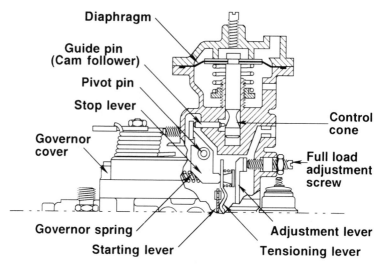

Fig. 6.69 The Bosch manifold pressure compensation device used on the VE pumps

surface, like a cam follower. Its outer end contacts the upper end of what is termed the *guide lever*, which is centrally pivoted. The lower end of this lever bears against the governor tensioning lever, so the load exerted by the governor spring and tensioning lever maintains the guide pin continuously in contact with the control cone.

When the boost pressure rises to a predetermined point, it pushes down the diaphragm, and with it the vertical pin, against the influence of its return spring. The guide lever, following the guide pin as it rides down the slope of the control cone, causes the starting lever to move in the direction that increases fuelling. In the event of failure of the turbocharger, the spring returns the diaphragm to the position at which the engine will operate in a naturally aspirated mode without smoke.

Another option is load-dependent injection timing. As load is increased at constant speed, increasing quantities of fuel have to be injected, so smoother operation can be obtained if the injection timing is simultaneously progressively advanced. The converse is true when load is decreased.

Retardation of injection timing with decreasing load is effected by introducing a modified governor shaft and sleeve, Fig. 6.70. As the speed rises, owing to a decrease in load, the governor sleeve slides towards the tensioning lever so, without the timing retardation facility, the transfer pressure would increase and therefore the injection timing mechanism would rotate the roller carrier to advance the timing. With the modified sleeve and shaft, however, the left-hand radial port in the sleeve, as viewed in the illustration, is uncovered and allows fuel to pass through it into the axial hole in the shaft and thence through the radial hole at its other end, to be bypassed to the inlet side of the pump. This causes a fall in the pressure in the pump casing and therefore also in the injection timing mechanism.

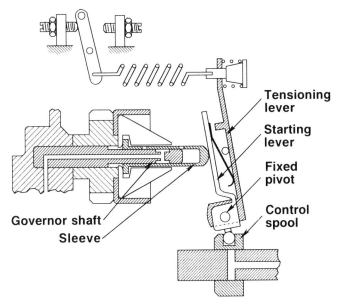

Fig. 6.70 Retardation of injection timing with decreasing load can be effected on the Bosch VE pumps by introducing a modified governor shaft and sleeve, as represented diagrammatically here

Consequently, the return spring in this mechanism causes the roller carrier to rotate to retard the injection.

If the load is increased and the speed therefore falls, the sleeve retracts until the radial hole is closed off. The resultant increase in pressure in the pump housing advances the injection again to the timing appropriate for the lighter load.

Several cold start injection advance actuation modules are offered by Bosch. The mechanical system is simply a cable-actuated lever, a projection from the stem of which extends into a slot in either the roller carrier ring or, in another version, into the timing piston. When the lever is moved about its pivot, it rotates the carrier, or slides the piston, into the enrichment position. In an automatic version, the lever is actuated by a temperature-sensitive expansion element, through which engine coolant is circulated. This unit, working on a principle similar to that of an engine coolant thermostat, has the advantage that the enrichment is always related to the actual coolant temperature. It can also be linked to a movable stop for limiting the travel of the engine speed control, to provide fast idling when the engine is cold.

What Bosch term a temperature dependent starting delivery control can be combined with the automatic cold start injection advance device. As the engine temperature rises, a stop on an extension of the cold start advance control lever is progressively moved over to a position in which it bars the cold start lever from being displaced the extra distance required for providing excess fuel for cold starting.

A hydraulically actuated cold start advance device can also be supplied. It comprises a modified pressure control valve, a pressure holding valve and an electrically heated expansion element, Fig. 6.71. Fuel pressure is communicated from the transfer pump to the bottom of the pressure control valve. Here, as in the standard pressure control valve, it acts on the piston lifting it against a pre-compressed coil spring above it. This piston does not lift far enough to uncover the radial ports through which fuel is released to the inlet side of the pump until, increasing with speed, the delivery from the pump has risen to the maximum level for which the unit has been designed.

Consequently, so long as the pressure holding valve is closed, the pressure under the piston, and therefore in the pump casing too, is always determined by the balance between pump delivery pressure (related to speed) and the compression in the spring above the piston, as it is with the standard pressure control valves.

What makes this different from the standard valve is that, drilled through the piston is an axial hole with a restrictor at its lower end, through which the fuel can pass into the spring chamber, and on down to the pressure holding valve. The bore of the restrictor is small so, if fuel is passing through it, the pressure above the piston is less than that below.

After the engine has started, the electrically heated element progressively

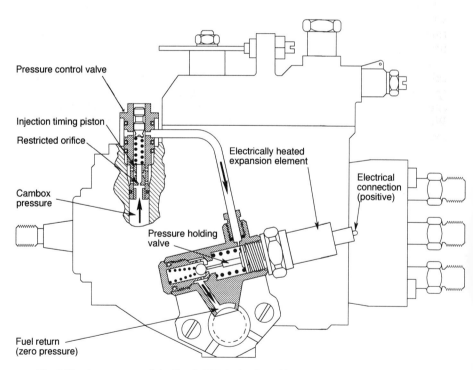

Fig. 6.71 Arrangement of the Bosch KSB hydraulic cold start injection advance device

DISTRIBUTOR TYPE INJECTION PUMPS

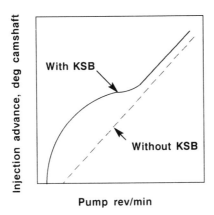

Fig. 6.72 Bosch illustrate the effect of the KSB is by this diagram

expands, ultimately lifting the ball in the pressure holding valve off of its seat, releasing the fuel to the inlet side of the pump. This causes the pressure above the pressure control valve piston to drop to zero. Consequently, the piston lifts, compressing its return spring further, fuel bleeds through the restrictor so the pressure below the piston, and therefore in the pump casing, settles at a lower level to advance the injection timing piston, for normal running. The effect on the pressure in the pump casing and automatic injection timing unit is illustrated in Fig 6.72.

Two fuel shut-off devices are available for stopping the engine. One is a cable-actuated lever that simply swings the starting lever over to open the spill ports. The alternative is a solenoid-actuated valve which, when the engine is switched on, is energised to open the inlet to the pump chamber. When the solenoid is switched off, a return spring closes the valve on to its seat again, so that no fuel can be delivered to that chamber.

Chapter 7

Combustion

Some aspects of the origins of the oil engine are not widely appreciated. The facts are that, in 1891, the Englishman Ackroyd-Stuart exhibited for the first time an engine designed to run on a fuel that was heavier than gasoline and was injected into the cylinders. This fuel was called gas oil, because it was used in the production of town gas. Two years later, in August 1893, Rudolf Diesel exhibited his first engine.

While both engines were based on the induction of air alone, the injection of the fuel towards the end of the compression stroke and compression ignition, there were fundamental differences between the two. Ackroyd-Stuart's engine ran at a relatively low compression ratio, and he used a heat-exchanger to preheat the air and thus assist both the vaporisation of the fuel and its ignition in the cylinder. He was the first to use mechanical injection of liquid fuel. Rudolf Diesel, on the other hand, adopted a higher compression ratio, to obtain compression ignition without the application of external heat, which is why he is accepted as the father of what is now called the diesel engine. However, he utilised an air blast to inject his fuel which, in the first instance, was powdered coal, but he soon changed to mechanical injection of liquid fuel and abandoned the air blast principle.

7.1 Thermodynamic characteristics of the diesel cycle

Because diesel combustion is compression induced and depends to a major extent on successful vaporisation and mixing over an extremely short time, its management presents much more complex and difficult problems than that of a spark ignition engine. The theoretical diesel cycle, Fig. 7.1, differs from the constant pressure cycle of the spark ignition engine, Fig. 7.2, as conceived by Otto. In reality, it tends to be close to the constant volume cycle, Fig. 7.1, but is actually a combination of the two, as shown in Fig 7.3, but with the corners rounded.

In all these cycles it is impracticable to extract work from the hot gases by expanding them right down to atmospheric pressure. This is because the exhaust valve must open before bottom dead centre, since some residual pressure is needed for scavenging the burnt gases from the cylinder in time for their replacement by incoming fresh gases. The work thus lost is represented by the shaded area in Fig. 7.1. Note that in the idealised diesel

COMBUSTION

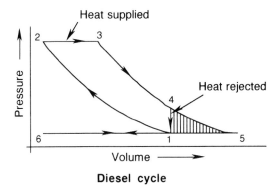

Fig. 7.1 Theoretical diesel thermodynamic cycle (combustion at constant pressure)

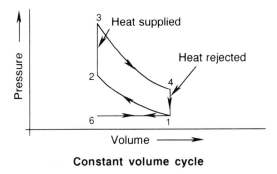

Fig. 7.2 Theoretical spark ignition thermodynamic (combustion at constant volume)

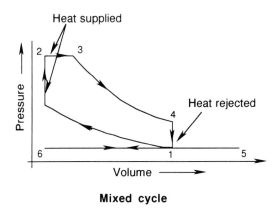

Fig. 7.3 In practice, the diesel cycle is approximately as shown here, but with the corners rounded, which amounts to a combination of the cycles illustrated in Figs. 7.1 and 7.2

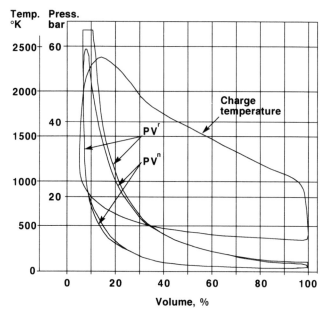

Fig. 7.4 Pressure-volume indicator diagrams with heat release diagram superimposed. The loop marked PV^r is based on the theoretical value of r (1.401) for adiabatic compression, while that marked PV^n is based on the more realistic value of 1.36. Actually, the value of n varies slightly with engine speed

cycle of Fig. 7.1, all the heat is assumed to be supplied while the piston is passing over tdc when, because upward motion of the piston has almost ceased, the gases are burnt at virtually constant volume. In practice, however, thes burning continues over part of the downward motion of the piston, during which the rate of expansion is such that the conditions approximate to constant pressure. An actual pressure–volume (indicator) diagram, together with the corresponding heat release diagram is shown in Fig. 7.4, and a pressure–crank angle diagram can be seen in Fig. 7.5.

Because engine speed determines the time available for complete combustion of the fuel, this is the parameter which, for any given compression ratio, primarily determines the optimum timing for the start of injection. With in-line injection pumps and mechanical governing, it is normally fixed at somewhere around 15 deg before top dead centre. However, with the distributor and rotary type pumps and electronic control it can be varied in relation to load and other parameters such as engine temperature. This has been covered in detail in Sections 6.13 et seq. The timing of the termination of injection, however, is in almost all instances varied in relation to the load on the engine. An exception is the GM unit injection system, Section 5.23.

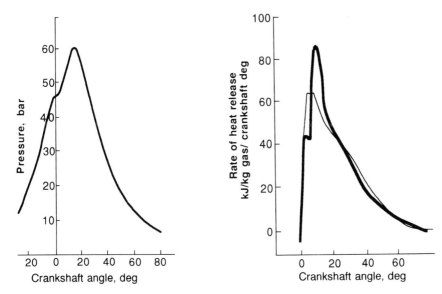

Fig. 7.5 Typical heat release and pressure diagrams plotted against crank angle. Here the heat release characteristics are plotted for 1600 rev/min (bold line) and 2800 rev/min (fine line) The kink at just above 45 bar in the compression curve is attributable to ignition delay

7.2 Differences between spark and compression ignition

In a spark ignition engine, a charge of air and fuel is of course supplied premixed to the cylinders and the source of combustion is a single point. In the diesel engine, on the other hand, air alone is initially supplied to the cylinders. It is then compressed until its temperature exceeds that necessary for spontaneous combustion of the fuel, which is only then injected into the cylinder as a very fine spray. Consequently, combustion is initiated at a number of points throughout the charge. The auto-ignition point of diesel fuel can be as low as 220 deg C, as compared with the ultimate compression temperature of about 600 deg C throughout most of the charge. Owing to heat losses, temperatures adjacent to the cylinder walls are significantly lower than the latter figure.

Since the droplets in the spray have to mix thoroughly with the air, evaporate, and then burn completely in the very short space of time remaining in the engine cycle, the primary requirement is that the fuel be consistently of a high quality, with properties rendering it suitable specifically for diesel engines. This is especially relevant for starting in very cold conditions, when the temperature on completion of compression can be as low as or even lower than 400 deg C, and the auto-ignition temperature as high as about 450–500 deg C. This is why some diesel engines require glow plugs to facilitate starting. When auto-ignition has begun, a small flame, or flames, may be alternately initiated and quenched, the temperature at which combustion spreads generally being between approximately 500 and 600 deg C.

Whereas, in a gasoline engine, control over power output is exercised by throttling the supply of mixture to the cylinders, in a diesel engine it is effected by regulating the quantity of fuel delivered through the injector nozzles. If more fuel is supplied than can combine with the oxygen available, the hydrogen content burns preferentially, so an extremely unpleasant black smoke is emitted from the exhaust. Therefore, the maximum power output has to be set by limiting the maximum quantity of fuel supplied per stroke to a level just below that at which black smoke is emitted.

Since combustion is initiated simultaneously at a number of centres distributed throughout the fuel air mixture, detonation as previously described in connection with gasoline engines, is not a problem. However, because of the very high air:fuel ratio under idling and light load conditions, when only minute quantities of fuel are being injected into what is, by comparison, a huge quantity of of air in the cylinders, an explosive combustion termed *diesel knock* may be heard. The causes of this phenomenon, and why in this context cetane number is important, are dealt with in Sections 2.3 and 7.4 respectively.

7.3 Mixture preparation and ignition

As has already been indicated, following the induction stroke, the charge of air is compressed until its temperature exceeds that necessary for auto-ignition of the fuel. Hitherto, the fuel has generally been injected into the cylinder about 15 deg before tdc, at pressures ranging from about 300 bar for indirect to 1000 bar for direct injection engines. The current trend, however, is towards later injection at pressures up to 1200 to 1300 bar, though as high as 1500 bar is being contemplated. This is for reducing the output of NO_x in the exhaust, Sections 8.1 and 8.2. Compression ratios range from about 14:1 to 24:1, typical values being 18:1 for direct injection engines and 22:1 for indirect injection engines, Section 7.5.

7.4 Ignition delay

The first point to note is that liquid diesel fuel will not burn: it has to be evaporated first. Obviously therefore some delay occurs from the outset, but this is not what is meant by the term ignition delay, which occurs even though the temperature of the air in the combustion chamber is above the auto-ignition temperature of the fuel vapour. Ignition delay is the interval between the evaporation and mixing of the fuel in the air and the initiation of combustion. It is generally of the order of 1000th of a second, but of course varies according to the properties of the fuel, size of droplets, their rate of mixing with the air, and the temperature. In principle, the delay can

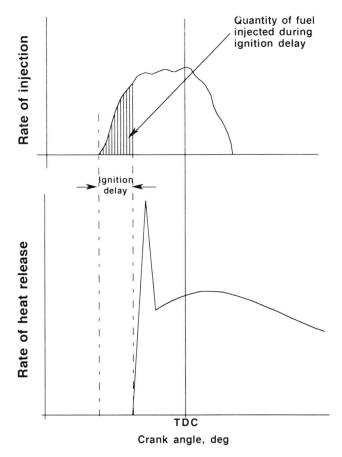

Fig. 7.6 Showing in detail how heat is released following ignition delay

be reduced by increasing turbulence, though of course there are practical limits to the degree of turbulence that can be accepted and indeed is effective.

During the delay period, mixing continues and pre-flame reactions take place, in which free radicals and aldehydes are formed. Until the visible flame appears, the curve of pressure against crank angle follows the line that it would have taken if the engine had been motored without fuel injection. Subsequently, the heat of combustion causes the pressure to rise rapidly, Fig. 7.6, and then falls again as the piston begins to descend on the power stroke.

For any given engine, fuel quality and compression temperature, the delay period is constant in terms of time, and therefore, as the speed of the engine increases, it becomes longer in terms of crank angle. Consequently, it is desirable, though not in all instances deemed practicable, to advance the injection timing with increasing speed.

The longer the delay period, the steeper will be the subsequent pressure rise and therefore the noisier the engine. This noise, diesel knock, is particularly loud when the engine is running under very light load or idling, especially after starting from cold. It becomes louder as the volatility of the fuel is increased and the cetane No. reduced.

The main reason for the increased noise when starting from cold is that, under these conditions, temperatures in the combustion chamber are low so the ignition delay is long, with the result that a higher proportion of the total fuel charge is injected before combustion starts. Also, only minute quantities of fuel are being injected, so it may not be well atomised, and the quantity of air available for combustion is, by comparison with that of the fuel, huge. The outcome is a sudden late release, at or near top dead centre, of a fairly large proportion of the total energy supplied and, therefore, the explosive combustion previously described as diesel knock.

7.5 Cold starting

To ignite the air–fuel mixture, its temperature must be above a minimum of at least 220 deg C which, for three reasons, is difficult to attain in very cold ambient conditions. First, the rate of heat loss from charge to the cylinder and combustion chamber walls may be considerable. This loss is particularly significant in indirect injection engines, owing to their high rates of swirl and losses during the rapid flow of the air through the throat. Secondly, because the oil film may not be intact on the cylinder walls, the rate of leakage past the piston rings may be high and the effective compression ratio therefore significantly reduced. Thirdly, the electrical output from the battery will be reduced because it is cold, and therefore the rate at which the starter motor rotates the engine will be correspondingly lower, allowing more time per stroke for leakage of the charge past the rings.

The smaller the engine the higher is the ratio of surface area to volume. Consequently, small direct injection engines may require a compression ratio of about 15:1 for starting from cold, while large ones may need only about 12.5:1. Indirect injection engines, on the other hand need even higher compression ratios of, at the very least 18:1. In practice, because engines produced in large quantities and exported have to be capable of operation in the coldest of climates, their compression ratios are generally set at significantly higher figures than those just quoted, see Section 7.13. However, there is a limit because, the higher the compression ratio the greater are the friction losses and loading on the bearings, pistons, crankshaft etc.

Several measures can be taken to alleviate the problems of cold starting. The incoming air can be heated or, for indirect injection engines, a glow plug can be installed in the swirl chamber. Extra fuel is usually injected under cold starting conditions, to compensate for condensation in the combustion chamber and leakage past the pumping elements which, under these conditions, are moving relatively slowly. Because the quantity of fuel needed for starting is so small, this leakage can constitute a significant proportion of the total. With electronic control, it is possible to advance the start of injection for cold starting, to compensate for the increase in ignition delay and to ensure that the main ignition phase occurs over top dead centre, Section 6.14.

7.6 Cold starting aids

For each engine, there is a critical temperature below which it will not start without assistance. Naturally, the higher the compression ratio, the lower is the critical ambient temperature. Increasing the compression ratio, however, can lead to rough running at normal temperatures. Because of the heat lost to the cold structure during the passage of the air into its swirl chamber, the indirect injection engine is the most likely to need assistance to get it started.

Traditionally, a glow plug has been used for raising the temperature above the critical level locally in the combustion chamber, by evaporating the fuel deposited on and in close proximity to it. Prior to about 1980, glow plugs generally had bare heating elements of the hairpin, or coiled hot wire type, projecting into the combustion chamber. However, the time taken for such devices to provide enough heat to start the engine was 20–30 sec or, in very cold conditions, even longer. Moreover, in the corrosive environment in the combustion chamber, their lives tended to be too short, so manufacturers developed much larger coiled heating elements enclosed in gas-tight steel glow-tubes. A modern example is the Lucas CAV Micronova, Fig. 7.7, which attains 850 deg C within 2 sec and which is installed in conjunction with a control for regulating the length of time that the complete set of glow plugs is energised.

Some of the fuel injected into and swirling around with the air in the combustion chamber is evaporated and ignited by the hot surface of the sheath, so that combustion can be initiated. To sustain this combustion, the glow plug is kept in operation for a few more seconds, dependent on ambient temperature, by the electronic controller. The efficiency of this device can

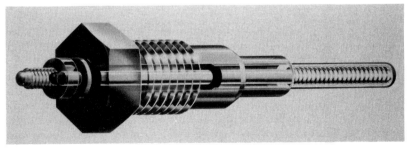

Fig. 7.7 The electrically heated element of the Lucas Micronova glow plug can attain a temperature of 850°C within 2 sec

be as high as 90%, as compared with about 35% for the traditional filament type glow plug, and its sheathed filament is well protected from erosion and corrosion. Only a single pole electrical connection is required and, since all the plugs can be connected in parallel, installation is simple.

Bosch produce a glow plug in which two coiled filaments connected in series are embedded in a refractory material (magnesia) inside the tubular sheath. The first coil, has a positive temperature coefficient (resistance increasing with temperature). This regulates the current flowing through the second coil, in the tip, thus preventing them from overheating, and ensuring that the drain on the battery is kept to a minimum.

Most modern glow plugs, with their high performance metallic filaments, generate heat rapidly enough to ensure starting from cold within 4–10 sec and, in extremely low temperatures, 15–20 sec. In some applications, the driver turns the ignition key a few degrees and then waits until a lamp on the dash fascia is illuminated to indicate that the glow plug is hot enough for him to rotate it further to switch on the starter motor. Another type of system is controlled by a thermo-time switch and relay: when the ignition key is turned, the lamp is immediately illuminated, but the engine is not ready for starting until it goes out again.

It would appear that the ultimate will be the system in which, when the ignition key is turned, all that follows is automatic, as with spark ignition engines. The benefits of the substitution of an electronic control for the bimetal thermo-switch formerly used are: greater precision, the heating element is energised only as long as it is actually needed, the risk of smoke

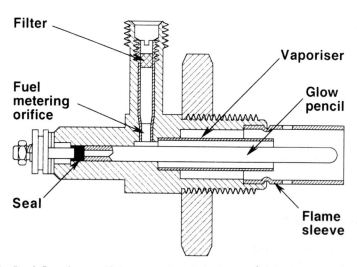

Fig. 7.8 Bosch flame heater. After entering through the filter at the top, the fuel passes around the glow pencil inside the vaporiser tube. The is vapour ignited as it passes over the tip of the glow pencil, which is at a temperature of about 1000°C, and the burning gases emerge into the induction manifold where combustion is completed

generation during cold idle can be greatly reduced, and the glow plug is automatically protected against overload.

Large direct injection engines need a greater heat input than that which can be supplied by either the filament or the sheathed element glow plugs. One solution to this problem is to heat the air before it enters the engine. This can be done by interposing an electrically heated tube or flange between the air intake and the induction ports. However, because its energy consumption is so high, it is suitable for engines of only up to, at the most, about 3 litres capacity.

A better alternative is to insert a flame heater in the induction manifold. This comprises a sheathed glow plug, Fig. 7.8, to which is piped fuel from either the supply line or the low pressure side of the pump. As the fuel passes over the surface of the heating element, which may be at a temperature of about 1000 deg C, it evaporates and ignites in the space between the sheath and a vaporiser tube around it, before mingling with, and thus heating, the ingoing air. Since, under starting conditions, the rate of induction of air greatly exceeds that required for combustion of the fuel injected into the cylinders, the quantity of oxygen consumed by the flame of the heater is of no practical significance.

7.7 The three phases of normal combustion

When the engine is running normally, combustion proceeds in three phases. As indicated in Section 7.1, the source of energy for igniting the mixture is the high temperature of the air that has been compressed in the cylinder. During the first phase, immediately following injection, the droplets of fuel tend to break up more finely and some evaporation occurs.

The second phase begins when flames appear. Since these, as previously indicated, are initiated simultaneously at a number of centres distributed throughout the combustion chamber, detonation as described in connection with gasoline engines, in Vol. 1, is not a problem. This phase is characterised by a rapid rise in pressure, which is ultimately halted and reversed as the piston begins to descend. In the third phase, which is the period during which the piston is descending, the remainder of the fuel is injected, mixed with and evaporated into the air, and finally burnt.

7.8 Direct injection combustion chamber design

In general, most European manufacturers produce diesel engines having relatively small diameter bowl-in-piston combustion chambers calling for high rates of swirl, Fig. 7.9, which is widely described as sending the air to look for the fuel. On the other hand, in the USA, large diameter bowls in the piston and low angular velocities of swirl are favoured, Fig. 7.10. This

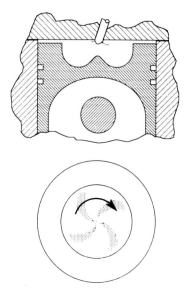

Fig. 7.9 Typically, in European diesel engines, the fuel is injected into a relatively small diameter combustion chamber in the piston crown. A high rate of swirl sends the air in search of the fuel.

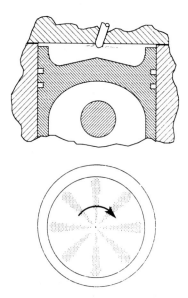

Fig. 7.10 American combustion chambers are generally of large diameter and swirl velocities relatively slow. The fuel is therefore injected at high pressure through multiple holes in search of the air

calls for higher injection pressures and is, in effect, sending the fuel to look for the air.

These contrasting design policies stem from different driving habits. In Europe, trucks are geared for, and drivers aim at, operation in the engine speed range giving maximum torque, optimum efficiency, and therefore minimum fuel consumption, in approximately the mid-speed range. The American driver, on the other hand, generally with longer distances to cover and low cost fuel, tends to keep his foot down and simply let the governor determine the speed, so trucks in that country are geared accordingly and, so far as practicable having regard to the fact that engine friction and pumping losses increase with speed, maximum efficiency is obtained higher up the speed range. Because higher pressures are needed for injection into larger diameter bowls, the US manufacturers tend to favour unit injection.

Combustion chambers of many different shapes are employed. Because of the need for a high compression ratio, the clearance between the crown of the piston and the flat lower face of the head, forming the top of the combustion chamber, is very small. Generally, the valves seat in rings of wear-resistant material inserted into the head casting and, to avoid having to machine recesses in the piston crowns to clear their heads at top dead centre, their stems are vertical.

The combustion chamber is a cavity in the piston crown, and usually coaxial with it, though not necessarily so. Reasons for eccentricity include a desire to avoid injecting on to the wall of the chamber where it is impracticable to install the injector vertically above its centre. Alternatively, the air swirl induced by the porting may not be coaxial with the cylinder, so the axis of the chamber is offset to coincide with that of the swirl. A disadvantage of an eccentric combustion chamber is that the temperature distribution over piston crown is not uniform, so thermal distortion may arise.

The cavities take many different forms, some of which are illustrated in Fig. 7.11. The simplest is the cylindrical combustion chamber. Incoming air is directed through the inlet port in a manner such as to cause it to swirl about the axis of the cylinder. As the piston rises during the compression stroke, the vortex is increasingly forced into the combustion chamber. Because this reduces its radius, while its energy content remains almost constant, its speed of rotation increases

At about 15 deg, or less, before tdc, injection begins and the fuel is swept around with the air. Further movement brings the piston very close to the flat base of the cylinder head, which causes the air between it and the flat portion of the crown of the piston to be violently displaced radially inwards. The result is a squish effect, the air spills over the edge of the combustion chamber, rolling down into it and generating eddies as it does so.

By modifying the shape of the combustion chamber, it is possible to change the nature of the eddies from what might be described as micro-turbulence to macro-turbulence, or even roll it over to swirl about a circular axis normal to that of the cylinder, thus forming a singe annular vortex, like a smoke ring, which is itself rotating bodily about the axis of the cylinder. A fine lip

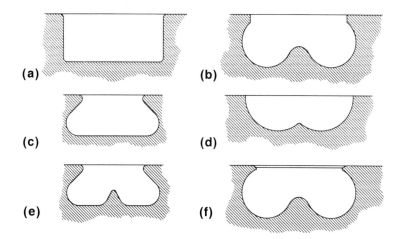

Fig. 7.11 A selection of combustion chamber shapes. The function of upward the projections in the centres of some is to locate the axis of swirl. Chambers (a) and (c) would tend to generate a high degree of random turbulence, that at (e) a combination of random turbulence in a toroidal vortex, those at (b) and (d) a high degree of macro and micro turbulence in a toroidal vortex, while that at (f) would tend to generate micro turbulence in a predominantly toroidal vortex

around the open edge of the chamber tends to retain liquid fuel in it, while also accentuating the vortex formation. If the lip is large, it tends to roll the squish, in the form of the previously mentioned annular vortex, into the chamber. A relevant point is that a lip of very fine section will tend to become hotter than the mass of metal that surrounds it and, if too hot, could soften and therefore be easily eroded.

Some chambers have flat bottoms, while others are raised in the centre to approach, to a greater or lesser degree, a toroidal section. The latter tend to encourage the generation of an annular vortex, while centralising the axis of swirl within the chamber. In some instances, there is little more than a small conical projection rising from the centre of the floor of the chamber, its function being solely to locate and hold the axis of swirl.

Particularly interesting is the Perkins Quadram combustion chamber used on their Phaser engine. This is of toroidal section but has vertical grooves at 90 deg intervals around its periphery, Fig. 7.12. The incoming air is directed tangentially into the cylinder and ultimately swirls around inside the combustion chamber. As it traverses the vertical grooves, it is broken up into eddies. Claims made for this system include a 13% gain in power output, 17% in torque, 8% improvement in fuel consumption and a reduction in peak pressure of 10% owing to a reduction in the ignition delay period, Fig. 7.13 and Section 7.4. The latter feature reduces mechanical loading, NO_x, and noise.

A benefit that had not been anticipated was that the clearance between the crown of the piston at tdc and the cylinder head was much less critical: varying the gap between 0.18 and 0.36 mm was found to have no measurable

COMBUSTION

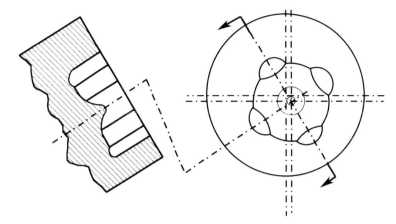

Fig. 7.12 Plan and section of the Perkins Quadram combustion chamber. It would appear that the vertical grooves in its periphery generate primarily macro, but also some micro, turbulence rotating about a vertical axis while the sharp upper edges of the chamber generate both micro and macro turbulence rotating about horizontal axes. All this turbulence is superimposed on a swirl generated by directing the air tangentially into the cylinder

effect on engine performance. Consequently, it was possible to produce the piston by diecasting and without machining the combustion chamber or crown. This avoided the problems that could have arisen because machining tolerances tend to drift out of tolerance limits. Also, as regards height, the pistons needed to be divided into only two grades for selective assembly.

7.9 Injection viewed in detail

In Europe, direct injection pressures of about 450–800 bar at the nozzles have been common though, to reduce both particulate and NO_x emissions, these have recently tended to increase to about 1200 bar. Indeed, pre-pressurised systems for delivery up to 1500 bar are under development. This pressure has been claimed as the practical limit because, beyond it, the fuel particle velocities would in any case exceed the speed of sound. However, not all authorities agree with this hypothesis. In the USA, for the reasons given earlier, the pressures have tended for many years to be higher, at about 1000 bar.

Looking at the fuel injection spray pattern in detail, Fig. 7.14, we find that it initially comprises a central core of liquid fuel at a density of about 840 g/m^3, surrounded by two distinctly different strata. Of these two, the innermost layer is created by the transfer of momentum from the central core to the air immediately surrounding it, and comprises fine globules of fuel intermixing with the air. The outer layer is a mixture of globules of fuel evaporating progressively into the air as they travel further from both the

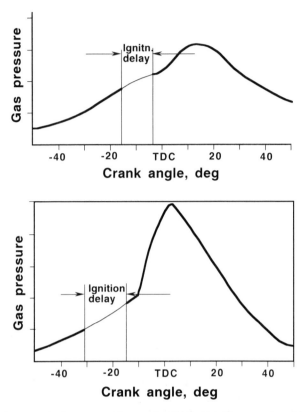

Fig. 7.13 Above, a cylinder pressure diagram for the Perkins Phaser engine with the Quadram combustion chamber, compared with, below, a similar diagram for a comparable engine with a conventional combustion chamber. With the Phaser combustion chamber, the ignition delay is shorter and the rise to peak pressure more gentle, generating less noise and lower mechanical loading

nozzle and the axis of injection. In other words, the fuel is looking for air with which to mix.

Clearly, the penetration of the spray in a quiescent chamber depends upon the momentum of the droplets and the drag force. The momentum of the leading droplets is boosted by a shunting effect of those following. Primarily the mass and velocity of the central core of liquid fuel determine the penetration of the jet. Given constant injection pressures, the smaller the orifice, the shorter is the penetration. Conversely, if the injection pressure is increased, the size of the orifice must be reduced if deposition of fuel on the walls of the combustion chamber is to be limited.

Provided mixing is good, flame is generally initiated at about 66% to 75% of the distance along the jet from the nozzle. It spreads rapidly downstream, but much more slowly upstream of the jet. Because the incoming fuel quenches it, the flame cannot actually reach the nozzle until injection has terminated,

COMBUSTION 199

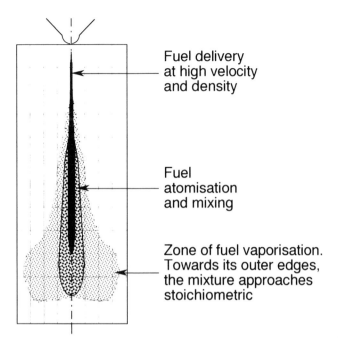

Fig. 7.14 *Typical pattern of fuel distribution in the jet issuing from an injector hole*

which helps to prevent carbon build-up around the hole. If mixing is poor, flame may be initiated only sporadically and immediately quenched. In extreme circumstances, it may fail to propagate until after injection has ceased.

As previously indicated, noisy combustion is mostly due to too large a volume of fuel being introduced and mixed with air during the delay period between injection and ignition. The result is too steep a rise in pressure when combustion ultimately occurs. To avoid this, the injector may be designed to function in two stages. Basically, there are two commonly used methods of doing this: either the initial injection can be at a lower pressure than the main injection, or injection can be totally cut-off between the two stages. The former is termed *dual injection* and the latter *split injection*. Arrangements for achieving both of these ends are described in Section 5.5, but here we are concerned only with the effects on combustion.

The initial, or pilot, injection is at a lower pressure than the main injection, so its penetration is shorter with the result that, without swirl, the main injection shot passes through the pilot pre-combustion zone. Consequently, with dual injection, a significant proportion of the fuel is ignited during the initial phase, but its combustion is then quenched by the main injection. With split injection, however, the positive cut-off between the pilot and main injection phases allows more time for pre-combustion before the main injection arrives and the tendency to quench it is reduced.

By injecting initially though one hole and directing the main injection shot through another along a different axis, it is possible to direct the pilot injection into the warmest area of the combustion chamber where it will assist combustion of the main shot more effectively than if both were to be coaxial. A major benefit from any pattern of two-stage injection is that the turbulence created locally by thermal expansion of the gases during the initial combustion and quenching phases, generates micro-turbulence which significantly enhances both the mixing of the main charge with air and its ultimate combustion.

7.10 Injector holes and spray penetration

Most manufacturers design injectors on the principle that three dimensional evaporation of fuel droplets in the combustion chamber is more effective than the two dimensional evaporation occurring after they have been deposited on the walls of the chamber. On this basis, the diameter of the nozzle orifice has to be such that the only a very small proportion of droplets injected at the maximum rate of fuelling penetrate as far as the opposite wall.

If the quantity of fuel required is large and the diameter of the combustion chamber small, multi-hole injectors are employed so that each hole can be small. Multi-hole injection has the advantage that it distributes the fuel more uniformly throughout the combustion chamber, but the holes may become more easily blocked. Tight production tolerances are needed on the diameters of such small holes and, because they call for higher injection pressures, erosion can be a problem. Moreover, multi-hole injection is impracticable for small engines requiring only minute quantities of fuel to be injected per cycle.

The higher the injection pressure the greater are the fuel droplet velocities and therefore the larger is the volume of fuel tending to impinge on the walls, while the shorter the duration of injection the less time there is for air to be entrained in the jet itself. Recently, the trend towards increasing injection pressures and shorter injection times, for reducing NO_x emissions, have been arousing increasing interest in the promotion of air entrainment from wall flows.

For many years MAN have been deliberately injecting the fuel on to the wall of the combustion chamber of their M series engines. The momentum of the fuel issuing from jet spatters it over a large area from which it can be evaporated rapidly by a high rate of swirl. Considerable eddying occurs in the boundary layer (between the stationary wall and high velocity flow adjacent to it) so the fuel is first swept up into the micro-turbulence thus formed, and then dragged out into the main flow. It would appear that the overall effect could be likened to that of two stage injection, with evaporation from the jet representing the primary and the slower evaporation from the wall of the combustion chamber the secondary injection. This could account for the relatively quiet combustion obtained with this system.

7.11 Mixing fuel and air

To mix the fuel and air thoroughly, either multi-hole injection is desirable or swirl and turbulence must be introduced into the air. To obtain the best possible results, however, both measures must be adopted. Considerable skill and a great deal of time is needed in development to optimise the ratio of the velocities of the incoming air and of swirl. Too high a rate of swirl will quench the flame, while too low a rate will fail to produce good mixing. Work continues on the development of computer models representing injection air flow, air entrainment by the jet, and evaporation and combustion of the fuel

In general, the aim has always been at producing a homogeneous mixture. This, however, is an ideal that is never achieved in practice, combustion being in most instances initiated at a number of points around the chamber and then rapidly spread by the swirl and turbulence. Basically, the swirl transports the jet of fuel as it issues from the hole in the nozzle converting it from an axial into a spiral stream inside the chamber. At the same time, it interacts with the jets to create more turbulence, thus mixing the two more thoroughly. Additional turbulence is created by both drag in the boundary layer adjacent to the wall of the chamber and the combustion itself, as well as by the squish described in Section 7.8.

7.12 The generation of swirl

Swirl is generated either in the inlet valve port itself or by *masking*, or *shrouding*, one side of the valve, or by a combination of both. The simplest measure is to keep the port relatively straight in the cylinder head and aiming it so that the air is directed tangentially relative to the cylinder wall. Further improvement can be had by arranging the port so that it spirals in towards the inlet valve seat, Fig. 7.15, from which it again issues tangentially relative to the cylinder wall. This sets the air flowing spirally even before it passes the valve seat. The significance of this is that the inertia forces tend to throw the air outwards in the spiral, so that a higher proportion of its mass entering the cylinder tangentially is closer to its walls than otherwise would be the case.

Masked valves are not so common on high speed engines because they tend to impede the flow of air into the cylinder and thus reduce volumetric efficiency. Among the serious disadvantages of masking valves are that they increase their mass, and therefore inertia forces, and they render the valve asymmetrical, which tends to result in thermal distortion leading to impairment of its sealing on its seat. Moreover, a great deal of engine development work is entailed and, in some configurations, the masks can cause more turbulence than swirl. Turbulence is of course generated more satisfactorily by other means, Section 7.8.

The masks are either welded to or, more commonly, forged integrally with

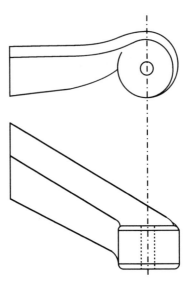

Fig. 7.15 A diagrammatic representation of an induction port designed to generate swirl

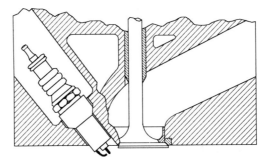

Fig. 7.16 Illustrating the form of a shroud integral with the inlet valve, in this instance in a spark ignition engine

the valves, Fig. 7.16. They extend part of the way around the periphery, Fig. 7.17, and rise to a level just upstream of the seats so that, when the valve is open, they virtually blank off part of the peripheral gap between the edge of the valve and its seat in the head. In this way, the stream of air, initially flowing tangentially into the cylinder, can be more positively turned into the direction of the required swirl about the axis of the cylinder. Depending on which side of the valve they are positioned, they can encourage either tumble or barrel swirl.

Barrel swirl is rotation of the air about the axis of the cylinder, while tumble swirl is rotation about an axis normal to that of the cylinder. An advantage of tumble swirl for gasoline engines is that it breaks up into

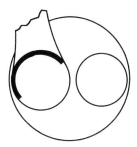

Fig. 7.17 Showing how the shroud in Fig 7.16 extends round the valve head to obstruct entry of the gas except in the direction that will generate swirl in the combustion chamber. The associated restriction of the inlet port area is of course a disadvantage

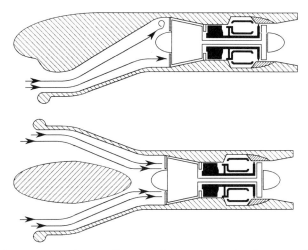

Fig. 7.18 Above, in a gas turbine an offset air intake generates swirl at the entry to the compressor whereas, below, a twin side-by-side entry does not

random turbulence as the piston comes up to top dead centre. However, barrel swirl is normally preferred for diesel engines, because it enables the movement of the air in their combustion chambers to be more closely controlled.

Incidentally, an aspect of valve port design that tends to be forgotten arises owing to the fact that, because of the need to align the outer end of the port with the induction pipe and to accommodate the actuation mechanism for the vertical valves, the axis of the port has to be turned through an angle which may approach 90 deg. This means that the air entering from the side tends to be carried over by its momentum to the opposite wall of the port, so the flow passing through the valve seat is far from uniformly distributed over its cross section. Aircraft engineers, observing a similar situation arising with fighter aircraft jet engine air intakes, Fig. 7.18, concluded that twin air

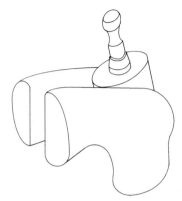

Fig. 7.19 Mechadyne, applying the principle illustrated in Fig. 7.18, have designed and patented a twin entry valve port for obtaining high efficiency of air flow into the cylinder

intakes, one from each side, are more efficient than a single intake from one side. On this basis, Mechadyne Ltd has designed and patented a porting arrangement with twin intakes, Figs. 7.19.

7.13 Indirect injection

In general, if the cylinder capacity of a direct injection engine is less than about 0.75 litres per cylinder, the diameter of the cylinders is so small that it is extremely difficult to avoid depositing a large proportion of the fuel on to the walls of the combustion chamber. To reduce the penetration of the fuel spray, the injector holes would have to be so small as not only to call for impracticably high precision in manufacture, but also to become fouled too easily and therefore cease to meter and mix the fuel properly. So difficult have these problems been to overcome that it was not until relatively recently (approximately a century after the invention of the diesel engine) that the first direct injection diesel engine having four cylinders each of 0.5 litres, the Perkins Prima, was put into quantity production.

Mostly, therefore, the small engines for cars still have indirect injection into small diameter spherical or cylindrical combustion chambers in the cylinder head. These are widely termed *swirl chambers*, though sometimes referred to as *pre-chambers*. The air stream, forced in from the cylinder during the compression stroke, is directed through a small diameter throat aligned tangentially relative to the wall of the combustion chamber, Fig. 7.20. This generates a very high velocity swirl, which tends to scrub off the walls any fuel that deposits there. The temperature of the charge in the chamber, and therefore its pressure, is rapidly increased by the combustion, forcing the still burning mixture and unburnt fuel back out through the throat into the main chamber. This outflow generates considerable turbulence, evaporating the remainder of the fuel, and burning it in the excess air present.

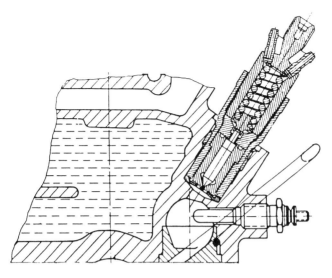

Fig. 7.20 The Lucas type LRC injector, which has superseded the Microjector for use with indirect injection combustion chambers. The alternative backleak connections facilitate installation. Not shown here is the duct for conducting the fuel down from the inlet at the top, through the body and nozzle, to the sac surrounding the lower end of the needle valve

For the following reasons, the thermal efficiency of the indirect injection is significantly lower than that of the direct injection engine, though still better than that of the spark ignition power unit. Because of the high ratio of surface area to volume of combustion chamber and throat of the IDI engine, the rate of heat loss to coolant is considerable, and it is aggravated by the high velocity of the swirl and of the flow of the gases through the throat. Moreover, the high compression ratios, of the order of 22:1 to 24:1 as compared with the average of about 18:1 for the large direct injection engines, increase not only the friction losses in the engine but also the mechanical loading on it. This lost energy of course cannot be turned into useful work.

The close proximity of the swirl chamber to the valve seats tends to affect adversely their cooling, and therefore can increase the wear and tear on the valve gear. Other disadvantages, include a tendency to overheat the piston crown locally where the jet of burning gases from the pre-chamber impinges upon it as it passes over top dead centre Also, there is some difficulty in mixing the fuel and excess air in the main chamber thoroughly enough to obtain optimum power output. The two last mentioned disadvantages, however, can be partially overcome by careful design, such as the incorporation of recesses in the piston crown, to bifurcate and deflect the flow from the nozzle and spread it over a large part of the piston crown and deflect it up into the cylinder, Fig. 7.21.

To reduce to a minimum the deposition of fuel on the walls of the small chambers of indirect injection engines, the fuel may be injected upstream

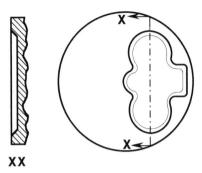

Fig. 7.21 Crown of a piston produced by AE Piston Products Ltd for an indirect injection engine

into the air entering the chamber through a narrow throat, so that the jet is shattered by the incoming air and the fuel evaporates as it is subsequently carried by it around the chamber. However, several alternative injection strategies are favoured by different manufacturers: for example, to distribute the fuel as uniformly as practicable throughout its volume, a single jet may be directed radially into the centre of the chamber or, as a compromise between uniform distribution and thorough mixing, along some axis between those of radial injection and into the incoming air stream.

Yet another arrangement is one in which, to facilitate cold starting, the jet is directed towards the hottest part of the combustion chamber, which may be as far downstream from the throat as about three quarters of the way around its periphery. Alternatively, a two-hole nozzle, Section 5.5, may be used in which one jet is directed towards the throat and the other towards the centre of the chamber.

7.14 Reducing heat loss from indirect injection combustion chambers

To increase the power potential by enabling the combustion chamber to withstand higher temperatures, and to reduce the thermal losses by reducing the rate of flow of heat through the walls, combustion chambers are sometimes made of refractory materials. A report on research done on this subject by Charleton, of Bath University and Sheppard *et al* of the Ford Motor Company has been published in the Proc. I.Mech.E, Vol. 205, No. D4, 1991.

Ceramic half-chambers replaced the Nimonic components originally used in the Ford 1.6 litre engine, Fig. 7.20, on which the work was done. The half-chamber insert, Fig. 7.22(b), was easier and less costly to manufacture than a complete chamber made entirely from Nimonic, and the cast iron of the head can cope with the levels of temperature experienced in the upper half. A ceramic half insert in a high grade steel sleeve was tried first because it could have been put into production with minimum changes to the assembly and machining lines. Moreover, it was particularly convenient for making direct comparisons between the performances of different materials.

COMBUSTION

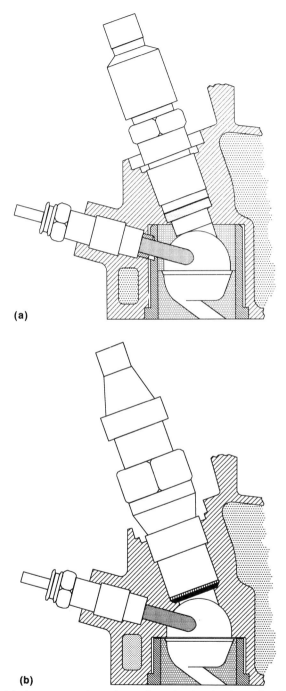

Fig. 7.22 At (a) is shown an experimental ceramic combustion chamber and, at (b) a half ceramic combustion chamber, both of which were used by Bath University in work done for the Ford Motor Company

Next, both halves of the chamber were made of ceramic and bonded together, and the steel sleeve was appropriately lengthened to encompass them, Fig. 7.22(a), and the head modified to accommodate them. This arrangement was more difficult to design and make, because of the need to locate accurately the injector and glow plug relative to not only each other but also the throat of the chamber. Sealing also presented problems.

Elsewhere, Isuzu and Kyocera have developed a silicon nitride chamber which is claimed not only to improve startability, but also to reduce HC emissions and low speed noise. The primary reason for its development, however, was to enable the engine to be uprated to a higher power output. Mazda introduced a chamber of similar material, for operation at temperatures up to 1000 deg C. It was claimed that by virtue of the high temperatures of the gases in the combustion chamber at light loads, the soluble fraction of the particulate emissions was reduced.

A major problem with ceramic chambers is the sensitivity of the material to stress raisers and its weakness in tension. For this reason the ceramic is usually contained in a sleeve of high grade, heat resistant steel. This sleeve is assembled to the ceramic by heating it to about 300 deg C, and then shrinking it on to the ceramic component. Incidentally, gap of about 1 mm between the periphery of the sleeve and the cylinder head has been found to provide a more effective barrier to heat flow than does the ceramic material.

In the paper referred to in the first paragraph of this Section, the thermal strength factor of the ceramic is given as $\sigma k/\alpha E$, where σ is the maximum allowable stress, k the coefficient of thermal conductivity, α the coefficient of linear expansion and E Young's modulus of elasticity of the material. Several different ceramics (aluminium titanate, TZP, Syalon and RBSN) were compared. In general, those having the higher tensile strengths also have relatively high thermal conductivities, approaching those of metals. Heat loss was found to be higher with large diameter ceramic inserts, presumably because the thicker walled chambers had larger areas of contact with the air gaps between them and the cylinder heads.

The properties of engineering ceramics vary widely. For example, TZP is very susceptible to thermal stress and its radial growth with temperature is 90 μm, as compared with 7 μm for aluminium titanate. Moreover, aluminium titanate has a negative coefficient of thermal expansion from 20 to 500 deg C, above which it becomes positive.

In the work done by Ford and Bath University, no improvement in performance could be obtained with a ceramic chamber and air gap for additional insulation, nor was the the reduction of emissions significant enough to warrant further investigation though, with the half ceramic chamber, the NO_x was reduced by 20% at maximum power. However, the results of these tests contradicted those obtained by the Japanese, but this could have been because the emissions of the standard Ford production engine with a metal swirl chamber were in any case very low. More significant improvements might be obtained with aluminium heads, because of the greater difference between the conductivities of aluminium and ceramic.

7.15 The small direct injection engine design problem

Consider the problems with which the designers of the Prima engine (0.5 litres per cylinder) were faced. First, even at maximum power, the quantity of fuel to be injected per firing stroke is of the order of 29 mm^3, which is the equivalent of a spherical globule of fuel only about 3.2 mm diameter. At idling, it is no more than 4 mm^3, or a globule little more than 1 mm diameter. Furthermore, at 4500 rev/min, the rate of injection is 2780 times per minute and, since injection has to be completed within about 30 to 35 deg of crankshaft rotation, the time available for each injection is less than a thousandth of a second.

To inject the fuel in such a short time, high pressures are essential. This, together with the fact that the cylinders are so small, means that a proportion of the fuel will probably be sprayed on to the walls of the combustion chamber, instead of into the air with which it has to mix intimately before it can be burned. Consequently, the mixing process tends to be delayed.

7.16 Direct injection and spark ignition compared

For the purpose of comparison, we shall use as examples a range of comparable production engines, all installed in Rover Group cars. From the outset, the Prima direct injection diesel unit installed in the Rover cars was jointly developed by Rover and Perkins, for production in both naturally aspirated and turbocharged forms. It is a 1993 cm^3 four cylinder unit, developing 46 kW naturally aspirated and 59.5 kW turbocharged, both at 4500 rev/min. The corresponding maximum torques are respectively 122 and 154 Nm at 2500 rev/min, Fig. 7.23.

In the Maestro, for example, the gasoline options are the 1.3 and 1.6 litre units or the 2.0 litre injected engine for the MG version. Accepting that comparable power outputs from a diesel and a gasoline engine of the same size are impossible, it is most appropriate to compare the 2.0 litre turbocharged Prima with the 1.6 litre gasoline unit.

Given that this gasoline unit develops 63.4 kW at 5500 rev/min and its maximum torque is 131 Nm at 3500 rev/min, we see that the power output of the diesel engine is only 6% lower but at 18% lower speed, though its maximum torque is a significant 15% higher at 29% lower speed. To be competitive with the gasoline power unit, therefore, a diesel unit has to be turbocharged. Fortunately, diesel engines are much more suitable than a gasoline units for turbocharging.

All other things being equal, we could expect the turbocharged diesel version to have usefully better acceleration over much of the speed range but tailing off at the upper end. The maximum speed, however, would be lower than that of the 1.6 litre gasoline powered version. As regards fuel consumption, on the other hand, the direct injection diesel engine scores hands down, since its efficiency is about 40% higher than that of the gasoline

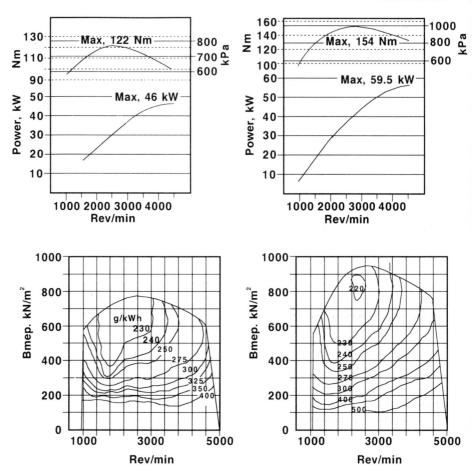

Fig. 7.23 Comparison of the performances of the Perkins Prima naturally aspirated (left) and the turbocharged (right) versions of the 2 litre direct injection engine

unit, which is a huge difference by any standards. Incidentally, even by comparison with an indirect injection diesel engine, the efficiency of the gasoline engine would still be about a 15% down.

It follows that, on grounds of good acceleration and low fuel consumption, especially in urban environments, a turbocharged diesel powered version can be particularly attractive. It does, however, tend to fall a little short as regards maximum speed. Although the diesel engine is significantly more costly than the gasoline power unit, the extra outlay is offset by the fuel savings within about 24 000 km (15 000 miles). Beyond this point, gains in terms of reduced fuel costs are made increasingly with mileage.

7.17 Direct injection, the ultimate aim?

Despite all the advantages of indirect injection systems, they also have the disadvantages outlined in Section 7.13. These not only reduce their efficiency in terms of conversion of the heat energy of the fuel into useful work but also, especially because a significant proportion of the heat generated by compression is lost to coolant, tend to make them difficult to start in cold weather. Because the compression ratio therefore may have to be 22:1 or even more, the indirect injection engine is noisy and the loads on its pistons, bearings, crankshaft and valve gear are more severe than with direct injection.

Another feature that contributes to an increased rate of wear is that, at high power outputs, the mixture in the swirl chamber is inevitably rich. Consequently, the fuel is not completely burned and the carbon particle content of the products of combustion is high, tending to contaminate the lubricating oil. For this reason, the oil change intervals with an indirect injection engine are shorter than for the direct injection and the service life of this type of engine is generally shorter than that of the direct injection type.

The direct injection engine, therefore, is potentially quieter, except during idling and at very light load; it is also more fuel-efficient, has a longer life and requires less maintenance, so its whole-life cost is lower. Yet another factor that makes it less costly is that its cylinder head casting is simpler because it does not have to contain a pre-chamber.

Furthermore, because of the inherently higher thermal and mechanical efficiency of the direct injection engine, the potential for increasing its power output per litre and per unit of weight is greater. As regards fuel consumption, the indirect injection engine is, as mentioned previously, about 15% better than a comparable gasoline engine but the direct injection type is potentially 25% better still. Turbocharging, by using some of the energy that would otherwise go to waste out of the exhaust, makes a significant contribution to overall efficiency, especially if accompanied by after-cooling.

Chapter 8
Exhaust emissions

Diesel and gasoline engines produce the same emissions, principally carbon monoxide (CO), carbon dioxide (CO_2), unburnt hydrocarbons (HC) and oxides of nitrogen (NO_x), but in very different proportions. On the other hand, owing to the low volatility of diesel fuel relative to that of gasoline and that carburettors are not employed, evaporative emissions are not of such significance, except in cold conditions. Crankcase emissions, too, are of less importance, since only pure air is compressed in the cylinder and blow-by constitutes only a minute proportion of the total combustion gases produced during the expansion stroke.

The background to emissions, as regards significance and toxicity, is covered in detail in Vol. 1, Chapter 13. Incidentally, it is of interest that in an interview with Ward's Engine and Vehicle Technology Update, 1 December 1990, R.C. Stempel, Chairman of GM, estimated the cost to the American economy of the control of emissions as required by the 1990 Clean Air Act, as 2.3% of the Gross National Product.

Each kg of fuel when completely burnt becomes 3.1 kg of CO_2 and 1.3 kg of H_2O. There are of course only two ways of reducing the CO_2 content in the exhaust: one is by improving fuel economy and the other by reducing the carbon content of the fuel, which is difficult to achieve with distilled crude oil based fuels. However, the magnitude of the problem should be assessed against the facts that commercial vehicles in Europe produce only 0.55% of all the CO_2 emissions worldwide, and water vapour, for example, is responsible for more of the greenhouse effect than CO_2. According to Prof. Böttcher, of the University of Leiden, the annual percentage of CO_2 emitted by the dormant Mount Etna volcano is greater that that from all the power stations in Holland. Among the products of combustion that are emitted in only minute quantities, even relative to those of primary concern, are the aldehydes, which contribute to the distinctive smell of diesel engine exhaust gases. Some aldehydes have been said to be carcinogenic, but this has not been proved.

Diesel fuel tends to contain significantly more sulphur than does gasoline. This sulphur takes the form of some insoluble compounds and a small proportion combines with the water produced by the burning of HC, to form sulphuric acid (H_2SO_4). It adds significantly to the particulate and smoke problem. Furthermore, the nature of the diesel combustion process is such that it produces between 5 and 10 times more solid particles than in the combustion of gasoline. The soiling effect of diesel smoke, because of its chemical and ultra-fine particulate content, is about 7 times greater than that of coal smoke.

It has been suggested that some particulates are carcinogenic. Again, however, this has not been proved. Typical mean concentrations of particulates in exhaust gas are 1 mcg/m^3, yet laboratory experiments with rats using 500 mcg/m^3 have been inconclusive. Moreover, Dr Robin Philipp of the Department of Epidemiology and Public Health Medicine at the University of Bristol, investigating a high risk group, twice reported, 1981 and 1984, that he had been unable to demonstrate any carcinogenic effect.

Because diesel power output is governed by regulating the supply of fuel without throttling the air supply, there is virtually zero CO in the exhaust under normal running conditions, except as maximum power and torque are approached. The same applies to HC. However, because of the relatively low volatility of diesel fuel and the extremely short time available for evaporation, problems do arise also when the engine is very cold. Under idling or light load conditions, the latter situation is aggravated by the fact that the minute volumes of fuel injected per stroke are not so well atomised as when heavier loads call for larger volumes of fuel to be forced at very high pressures through the injector holes.

As a diesel engine is opened up towards maximum power and torque however, NO_x output increases because of the higher combustion temperatures and pressures. As mentioned in Vol. 1, NO_x output is a function of combustion temperature but also, to a lesser degree, on pressure and dwell period at that temperature and pressure.

8.1 Reduction of emissions — conflicting requirements

Measures taken to reduce NO_x tend to increase the quantity of particulates and HC in the exhaust, Fig. 8.1. This is primarily because, while NO_x is reduced by lowering the combustion temperature, both soot and HC are burned off by increasing it. In consequence, some of the regulations introduced in Europe have placed limits on the total output of HC and NO_x, instead of on each separately, leaving manufacturers free to obtain the best compromise between the two.

Because the reduction of NO_x has to be done in an oxygen-free (reducing) atmosphere, and there is mostly an excess of oxygen in diesel exhaust gases, the 3-way catalytic converter used for gasoline engines is impracticable. On the other hand, because of the presence of this excess over most of the operating range, cleansing the exhaust of CO and HC is not so difficult.

The problem of emission control, however, is not so severe as might be implied from the last paragraph. Both heat to exhaust and NO_x output become significant only as maximum torque and power are approached. At lighter loads, the gases tend to be cooled, not only because of the large expansion ratio of the diesel engine, but also by the excess air content. Since the proportion of excess air falls as the load increases, oxidising catalysts

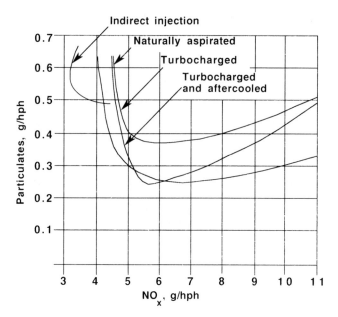

Fig. 8.1 US FTP cycle tests in the late 1980s demonstrated how reducing particulates entails an increase in NO_x

can be used without risk of overheating, even at maximum power output. Indeed, many diesel powered cars are equipped with oxidising catalytic converters.

As previously implied, the output of NO_x can be reduced most readily by lowering the peak combustion temperature. This can be done in several ways. By delaying the injection of fuel so that most of the combustion process occurs during downwards movement of the piston on the expansion stroke, the peak temperature of the charge is reduced. However, the associated increase in fuel consumption limits the degree of retardation that can be usefully applied. Water injection has also been used to cool the combustion process, but this introduces complexity into the injection system and other problems outlined in the next Section. Yet another way of reducing oxygen availability to limit NO_x output is to lower the rate of swirl, but this increases specific fuel consumption. For turbocharged engines, direct cooling of the charge by passing it through a heat exchanger is effective but costly. Exhaust gas recirculation (EGR) through the combustion chamber lowers the combustion temperature, by diluting the charge and thus limiting the quantity of oxygen available to the individual fuel droplets. However, it too introduces corrosion and wear problems detailed in Section 8.2.

Fuel blending and quality has a profound effect on emissions. Since fuel properties and qualities are interrelated, it is difficult to vary one property unilaterally. Indeed, efforts to reduce one exhaust pollutant can increase others and adversely affect other properties. More on this subject is to be found in Sections 8.2 and 8.8–11.

INDIVIDUAL EMISSIONS IN DETAIL

8.2 Oxides of nitrogen, NO_x

Air comprises 80% nitrogen, so despite the relatively inert nature of this gas, it is not entirely surprising that, at high temperatures and pressures, some oxides of nitrogen are produced in the combustion chamber. Actually, there is a critical temperature, which varies around 1370 deg C according to the pressure, below which oxides of nitrogen do not form in significant quantities. This is why, to eliminate NO_x, it is necessary only to reduce peak combustion temperature. The term NO_x is used to cover all the oxides of nitrogen emitted, which include mainly NO and NO_2 of which, as explained in detail in Vol. 1, Section 13.5, NO_2 is the most significant as regards its adverse effects.

To understand the effects of fuel properties on NO_x output, certain basic facts must be born in mind. First, its output depends on not only the peak temperature of combustion but also the rate of rise to it and fall off from it. Secondly, the combustion temperatures depend on primarily the quantity and, to a lesser degree, the cetane No. of the fuel injected.

Increasing the cetane No. has two effects, one of which is favourable and the other unfavourable for the formation of NO_x, and therefore has little effect on NO_x output. This is explained as follows. Increasing the cetane No. reduces the delay period, so the fuel starts to burn earlier. Consequently, higher temperatures and therefore more NO_x are generated, while the burning gas is still being compressed, Fig. 8.2. The second effect is a reduction in the quantity of fuel injected before combustion begins. This, by reducing the amount of fuel subjected to uncontrolled burning, reduces the peak combustion temperature. The net result of the two effects, is relatively little or even, in some circumstances, no change in NO_x output An interesting feature in Fig. 8.3, is the enormous difference between the NO_x outputs from direct and indirect injection systems.

Increasing fuel volatility is widely believed to reduce NO_x. This, however, is in illusion: what happens in reality is that the weight of fuel injected is reduced, and the engine is therefore derated. Consequently, combustion temperatures are lowered. This is explained in more detail in Section 8.9, in connection with black smoke.

As previously mentioned, because the diesel engine combustion is unsatisfactorily with a stoichiometric mixture, a 3-way converter comprising a reducing catalyst followed by an oxidising catalyst cannot be used. At the time of writing, the overall output of NO_x from the diesel engine is, on average, between 5 and 10 times that of equivalent gasoline power unit with a catalytic converter, but it obviously will be reduced as diesel combustion control techniques improve. Without catalytic conversion, however, the peak output of NO_x from the gasoline engine exhaust tailpipe, owing to higher peak combustion temperatures, can be more than double that from a diesel unit.

Efforts are being made to develop catalysts suitable for diesel application, but no satisfactory solution has been found so far. Work with the zeolite

216 EXHAUST EMISSIONS

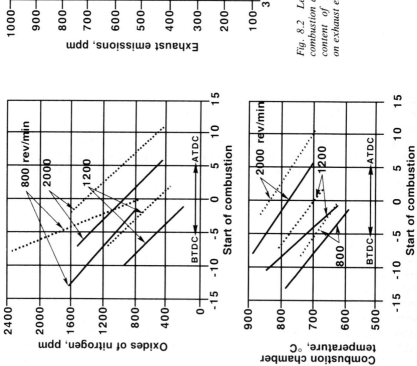

Fig. 8.2 Left, tests done by BP have shown how the timing of combustion affects the engine temperature and therefore the NO_x content of the exhaust gas; right, the effects of cetane No. on exhaust emissions in general

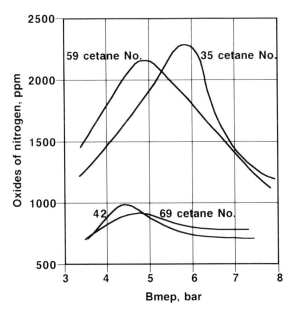

Fig. 8.3 Effect of cetane No. on NO_x in the exhaust with direct and indirect injection

catalyst Cu-ZSM-5, which looks promising, has been described by Iwamoto and Izuno in *Proc.I.Mech. E, 207 D1 1993*, but problems of durability, corrosion and overheating remain to be solved.

Unfortunately, most of the currently conventional methods of reducing NO_x also reduce efficiency and therefore increase fuel consumption and the output of CO_2. The relationship between NO_x output and fuel consumption is illustrated in Fig. 8.4. In general, NO_x tends to form most readily in fuel-lean zones around the injection spray.

As indicated in Chapter 7, to reduce NO_x by cooling the charge, either water injection or exhaust gas recirculation (EGR) can be employed. Water injection cools the charge, and the heat absorbed by the water converts it into steam which, together with the gaseous products of combustion, does work on the piston during the expansion stroke. However, it entails the installation of additional costly injection equipment and introduces corrosion problems. Exhaust gas recirculation reduces the power output and causes both corrosion and wear. Moreover, the associated reduction in the oxygen concentration in the combustion chamber, leads to smoke emission at high loads, which is why its use has to be confined to operation at moderate loads. Consequently, electronic control of EGR is essential, and therefore engine management systems are now being developed for satisfying future emissions regulations. Fortunately, however, heavy commercial vehicles are driven most of the time in the economical cruising range, maximum power and torque being needed mostly for brief periods.

Reduction of the rate of swirl is another way of reducing the output of

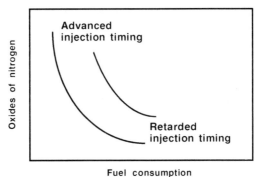

Fig. 8.4 Two curves showing the relationship between fuel consumption and NO_x in the exhaust of a turbocharged diesel engine, left with and right without charge cooling

NO_x. It increases the time required for the fuel to mix with the air, and therefore reduces the concentration of oxygen around the fuel droplets. Consequently, the temperature of combustion does not rise to such a high peak peak. However, it also reduces thermal efficiency. Moreover, unless measures are taken, such as increasing the number of holes in the injector nozzle and reducing their diameter, to shorten the lengths of the sprays, more fuel tends to be deposited on the combustion chamber walls. As explained in Chapter 7, a problem with small diameter holes is that they more easily tend to become blocked.

Delaying the start of injection has the effect of reducing peak temperatures because the combustion process builds up to its peak later in the cycle, when the piston is on its downward stroke and the gas is therefore being cooled by expansion. In these circumstances, to get a full charge of fuel into the cylinder in the time remaining for it to be completely burnt, higher injection pressures are needed. Therefore, to avoid increasing the proportion of fuel sprayed on to the combustion chamber walls, the holes in the injector must again be smaller in diameter and larger in number.

Turbocharging increases the temperature of combustion by increasing both the temperature and quantity of air entering the cylinder. Aftercooling, however, can help by removing the heat generated by both compression of the gas and conduction from the section of the turbocharger through which the hot exhaust gas passes. At the same time, by increasing the density of the charge, it improves thermal efficiency and power output. The net outcome of turbocharging with charge cooling, therefore, is generally an increase or, at worst, no reduction in thermal efficiency. Naturally, the higher pressures in the combustion chamber call for higher injection pressures to ensure that the fuel is adequately atomised.

8.3 Unburnt hydrocarbons, HC

As explained in Vol. 1, hydrocarbons in the outgoing gases are the result of incomplete combustion of not only the fuel but also lubricating oil. With a diesel engine, they arise in several circumstances, and are the principal cause of the unpleasant smell of diesel exhaust fumes.

The first is a mixture too lean for efficient burning. This failure to burn completely is especially liable to occur at low temperatures and light loads, including during the ignition delay period. In the latter circumstance, the pre-combustion processes are partially inhibited, leading to failure of some of the mixture to burn during the main combustion period.

Secondly, when starting and warming up from cold, the low volatility of diesel fuel relative to gasoline, and the short period of time available for it to evaporate before combustion starts, cause unburnt HC to become a major problem. In these circumstances, fuel droplets, together with water vapour produced by the burning of the hydrogen content of the remainder of the fuel, issue from the cold exhaust pipe in the form of what is generally termed *white smoke*, but which is in fact largely a mixture of fuel and water vapour. At about 10% load and rated speed, both HC and CO output are especially sensitive to fuel quality and, in particular, cetane No.

Thirdly, after cold starting and during warm-up, a higher than normal proportion of the injected fuel, failing to evaporate, is deposited on the combustion chamber walls. This further reduces the rate of evaporation of the fuel, so that it fails to be ignited before before the contents of the chamber have been cooled by expansion to a level such that ignition can no longer occur. Similarly, the cooling effect of the expansion stroke when the engine is operating at or near full load can quench combustion in fuel-rich zones of the mixture. This is the fourth potential cause of HC emissions.

Unburnt hydrocarbons tend, in any case, to become a problem at maximum power output, owing to the difficulty under these conditions of providing enough oxygen to burn all the fuel. As fuel delivery is increased, a critical limit is reached above which first the CO and then the HC output rise steeply. Injection systems are normally set so that fuelling does not rise up to this limit, though the CO can be dealt with by incorporating a catalytic converter in the exhaust system.

Another potential source is the contents of the sack volume in the injector, Section 5.3. Some of the fuel thus trapped may tend to evaporate into the cylinder after termination of combustion, when the injector needle has seated. Finally, the crevice areas, for example between the piston and cylinder walls above the top ring, also contain unburnt or quenched fractions of semi-burnt mixture. Expanding under the influence of the high temperatures due to combustion and falling pressures during the expansion stroke, and also forced out by the motions of the piston and rings, this HC content finds its way into the exhaust.

In general, therefore, the engine designer can reduce HC emissions in three ways. One is by increasing the compression ratio, the second is by increasing the specific loading by using a smaller engine for the type of operation, and the third by increasing the rate of swirl both to evaporate the fuel more rapidly and to bring more oxygen into intimate contact with it.

Reduction of lubricating oil consumption is another important aim as regards not only control of HC but also, and more importantly, particulate emissions. Whereas oil consumption at a rate of 1% of that of fuel has hitherto been regarded as the norm, the aim now is generally nearer to 0.2%.

Avoidance of cylinder bore distortion can play a significant part in the reduction of oil consumption. The piston rings tend to ride clear over and therefore fail to sweep the oil out of the pools that collect in the hollows formed by the distortion of the bores, thus reducing the effectiveness of oil control. Other means of reducing contamination by lubricating oil include improving the sealing around the inlet valve stems, the use of piston rings designed to exercise better control over the thickness of the oil film on the cylinder walls and, if the engine is turbocharged, reduction of leakage of oil from the turbocharger bearings into the incoming air.

8.4 Carbon monoxide, CO

Even at maximum power output, there is as much as 38% of excess air in the combustion chamber. However, even though CO should not be formed, it may in fact be found in small quantities in the exhaust. The reason is partly that, in local areas of the combustion chamber, most of the oxygen has been consumed before injection ceases. Fuel injected into these oxygen-starved areas cannot burn completely to CO_2.

8.5 Particulates and black smoke

Regulations define particulates as anything that is retained, at an exhaust gas temperature of 52 deg C, by a filter having specified properties. It therefore includes liquids as well as solids. Particle sizes range from 0.01 to 10 μm, the majority being well under 1.0 μm. While smoke comprises mainly carbon, the heavier particulates comprise ash and other substances, some combined with carbon. The proportions, however, depend on the type of engine.

Work done by Horrocks, of Ford, on two versions of their 1.8 litre engines, producing 0.096 g/km and 0.098 g/km respectively of particulates, revealed that, in both, the particulates comprised fuel and oil derived HC, sulphate+water, and insolubles, but in widely differing proportions, Table 8.1 (Proc I.Mech.E, Vol. 206, No. D4). Needham et al, of Ricardo, averaging the particulate content of the exhaust of 16 turbocharged and aftercooled heavy duty DI engines run on the US FTP test with the standard fuel, again showed different proportions (Proc I.Mech.E, Vol. 205, No. 3). In this case, the average particulate content of the gases was 0.37g/hp h (0.279 g kW h), the latter units being the most appropriate for the testing of heavy duty diesel engines, Section 9.2. Despite the differences in units, the percentage figures can, of course still be compared meaningfully, even bearing in mind that the conditions of operation differ.

A clear illustration of the relative proportions of the emissions and of the two main constituents of the particulates in the exhaust gas of a 1990s heavy duty diesel engine has been given by Volvo, Fig. 8.5.

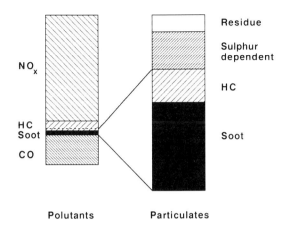

Fig. 8.5 Left, analysis of pollutants in the exhaust; right, analysis of the particulates only, the lower two components being a magnification of the HC and soot bands in the left-hand diagram

Measures appropriate for reducing the fuel and oil content of the particulates are the same as those already mentioned for reducing HC emissions, Section 8.3. The overall quantity of particulates can be reduced by increasing the injection pressure and reducing the size of the injector holes, to atomise the fuel better. This, however, tends to increase the NO_x content. Increasing the combustion temperature tends to burn the loose soot deposited on the combustion chamber walls. Various measures have been taken to increase the temperature of these particulates, though mostly only experimentally. They include insulation by introducing an air gap, or some other form of insulation, between the chamber and the remainder of the piston, and the incorporation of ceramic combustion chambers in the piston crowns.

Some particularly interesting work on combustion chamber insulation that has been done recently for Volvo is illustrated in Fig. 8.6. The ceramic insulation is about 5 mm thick around the ports and head, but significantly thicker on the crown of the piston, where the main problem is how to prevent it from cracking and being thrown off. Insulation increases combustion temperature and reduces losses to coolant so, by virtue of delayed injection and rapid combustion, it can possibly be made to have a beneficial effect as regards the generation of NO_x. However, the heat to exhaust is increased, which strengthens the case for turbo-compounding, Fig. 8.7.

To a lesser extent, insulation can be effected by the use of a slipper type piston: this has been done on a Volvo 10 litre engine. Because smaller clearances are practicable with such a piston, both noise levels and oil consumption are lowered, the latter reducing particulate emissions. Furthermore, the combustion chamber bowl keeps warmer, which helps to prevent the generation of smoke during idling.

Reduction of the sulphur content of the fuel also reduces particulates. Although the proportion of sulphate + water is shown in Table 8.1 as being

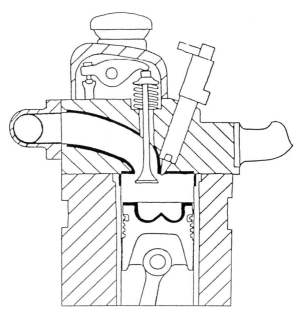

Fig. 8.6 Up to 5 mm lining of ceramic applied experimentally by Volvo to the exhaust port, and thicker in the combustion chamber, to reduce heat losses and thus to increase overall thermal efficiency

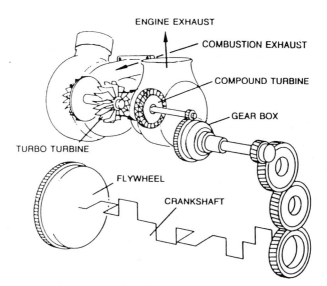

Fig. 8.7 With high temperature rapid burn and delayed injection timing at a high rate to reduce NO_x, compounding the diesel engine with a gas turbine can increase overall thermal efficiency to 50%, from the maximum of about 40 to 45% obtainable without compounding

EXHAUST EMISSIONS

Table 8.1 ANALYSES, EXPRESSED IN PERCENTAGES, OF PARTICULATES FROM DIFFERENT TYPES OF DIESEL ENGINE

Engine type	Fuel derived HC	Oil derived HC	Insoluble ash	Sulphates + water
Ford 1.8 DI	2	15	13	70
Ford 1.8 IDI	2	48	20	30
Average heavy duty TCA	14	7	25	44

Note: Horrocks differentiated between the carbon and other ash (at 41% and 13% respectively), making the total of 44%

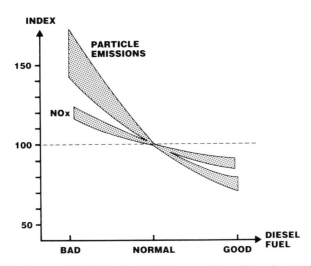

Fig. 8.8 Volvo have established this relationship between fuel quality and particulate emissions

only 2% of the total, if the insoluble sulphur compounds are added, the total becomes more like 25%. Because most measures taken to reduce NO_x increase particulates, the most appropriate solution is to use fuels of high quality, primarily having low sulphur and aromatic contents and high cetane number. The relationship between fuel quality and particulate and NO_x output has been demonstrated by Volvo, Fig. 8.8.

A small proportion of the particulates is ash, most of which comes from burning the lubricating oil. Reduction of sulphur in the fuel reduces the need for lubricating oils to contain substances that neutralise the acid products of combustion, and it is these additives that are largely responsible for the ash content.

Incidentally, sulphur compounds can also reduce the efficiency of catalytic converters for the oxidation of CO and HC. In so doing, they form hydrogen disulphide, which accounts for the unpleasant smell of the exhaust when fuels with high sulphur contents are burnt in an engine having an exhaust system equipped with a catalytic converter.

An ingenious method of reducing visible particulates emitted from a turbocharged engine in a bus has been investigated by MAN Nutzfahrzeuge, of Nuremberg. Compressed air from the vehicle braking system is injected in a controlled manner into the combustion chambers to burn off the carbon. This increases the exhaust gas energy content, and therefore compensates for turbocharger performance falling off under light load, including initially during acceleration and while gear changes are being made.

8.6 Particulate traps — general

Basically, particulate traps are filters, mostly catalytically coated to facilitate their regeneration (burning off of the particles that have collected). Many of the filters are extruded ceramic honeycomb type monoliths, though some are foamed ceramic tubes. The honeycomb ceramic monoliths in many instances differ from those used for catalytic conversion of the exhaust in gasoline engines: the passages through the honeycomb section are sealed alternately along their lengths with ceramic plugs, Fig. 8.9, so that the gas passes through their porous walls. These walls may be less than 0.5 mm thick.

Ceramic fibre wound on to perforated stainless steel tubes has also been used. These are sometimes termed candle or deep *bed* type filters. Their pores are larger than those of the honeycomb type, and their wall thicknesses greater.

If catalyst assisted, regeneration is done mostly at temperatures around 600 deg C, and if not, at 900 deg C. Alternatively, special catalysts may be used to lower further the ignition temperature of the particles. These are mostly platinum and palladium, which are useful also for burning HC and CO. Sulphate deposition adversely affects the platinum catalyst, but the formation of sulphates is largely inhibited by palladium. Copper, has been used too, because it reduces the carbon particulate ignition temperature to about 350 deg C but it suffers the disadvantage of a limited life. Better results have been obtained with alloys of copper with silver, vanadium and titanium. In other instances, fuel additives are used to reduce the burn-off temperature to as low as 200 deg C.

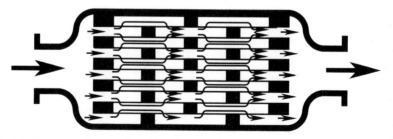

Fig. 8.9 Illustrating the principle of flow diversion through a particle trap

It is important to regenerate the filters reliably at the appropriate intervals, otherwise they will become overloaded with soot, which can then ignite and burn uncontrollably, developing excessive heat and destroying the filter. Even with normal burn off, however, local temperatures near the centre of the beds can be as high as 1200 deg C.

Some particulate filters rely solely on excess air in the exhaust for burning off the particulates. Others have extra air injected into them either from the brake system or by a compressor or blower. With air injection, burn-off can continue when the engine is operating at or near maximum power and therefore with little excess oxygen in the exhaust. At lighter engine loads, once ignition has been initiated, combustion can continue with the burner extinguished.

With some systems the engine must be stopped while regeneration is in progress but with others, regulated by electronic control, either the particulates are burnt off intermittently while the vehicle is running or there are two filters in parallel, with butterfly or deflector valves directing the exhaust gases first through one filter, while the other is being regenerated, and then *vice versa*. It is also possible to program the electronic controller to bring both filters into operation simultaneously as maximum power and torque output is approached, to cater for the increased exhaust flow under these conditions. This enables smaller filters to be used.

Claimed efficiencies of particulate removal range from about 70% to 95%, but some of these claims are suspect. However, since the carbon content is easiest to burn off, up to 99% of it can be removed. Currently, the useful lives of the catalytically coated filters vary enormously from type-to-type

Since we are in the early stages of development, the situation is in a continuous state of flux: some systems no doubt will soon fall by the wayside, while others will be further developed. Clearly, however, if particulates can be removed in the combustion chamber by using better quality fuel or taking measures to improve combustion, it will be to the advantage of all concerned except, of course, the manufacturers of particulate filters!

All filter systems are bulky and present burn-off problems, including high thermal loading. None produced so far can be regarded as entirely satisfactory, and none have been used for cars. Interestingly, VW have been investigating the possibility of using an iron-based additive to reduce the oxidation temperature of the deposits down to that of the exhaust gas. A Corning monolithic ceramic filter is used, and the additive, developed jointly with Pluto GmbH and Veba Oel, is carried under pressure in a special container. An electronic control releases additive automatically, as needed, into the fuel supply line, so that regeneration proceeds at temperatures as low as 200 deg C.

Of the systems currently available for commercial vehicles, those requiring the engine to be stopped during burn-off are unsuitable for any automotive applications other than city buses and large local delivery vehicles operated on regular schedules. The continuous and cyclic burn-off filters are generally even more bulky. On the other hand, many of these systems are claimed to serve also as a silencer.

8.7 Particulate traps in detail

One of the particulate traps in production at the time of writing, is Volvo's Cityfilter system, Fig. 8.10. It has a catalytically coated honeycomb type core, or filter, with an electric heater beneath. The exhaust gas passes in through the lower pipe connection, up through the honeycomb and out of the connection near the top, on the opposite side. During normal operation the electric heater is inoperative.

Sited together at the same level as the exhaust inlet connection are a third pipe and an electrical connection. These are for regeneration, for which the vehicle must return to base. The engine is switched off, purge air pumped into the third pipe, and a mains supply plugged into the electrical connection. The passage of air up through the heating element and on over the catalyst-treated core burns off the hot accumulated particulates. In normal operation of the vehicle, the catalyst is claimed to reduce the HC by 60% and CO content by 50%, while 80% of the particulates are trapped. The working life of the equipment is 300,000 km and its maximum range of operation, before regeneration of the core becomes essential, is 300 km.

Deutz Service Division of the KHD Group have produced what they call the DPFS filter system, Fig. 8.11, claimed to remove between 95% and 99% of the carbon particulates and about 70% of the remainder. It is about the size of a conventional silencer for the type of vehicle in which it is installed, and is claimed to be as least as effective as the silencer. Regeneration is automatically effected, by a diesel fuelled burner, at intervals of between 10 and 13 hours while the vehicle is running. As in most systems, the ceramic honeycomb core is housed in a stainless steel shell.

This unit is electronically controlled. When the back pressure generated

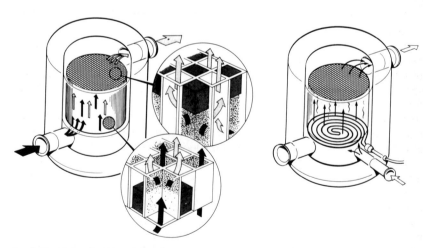

Fig. 8.10 Volvo Cityfilter: left, exhaust gas flow through the filter: right, regeneration by means of an external air supply and a mains powered the heating element

EXHAUST EMISSIONS 227

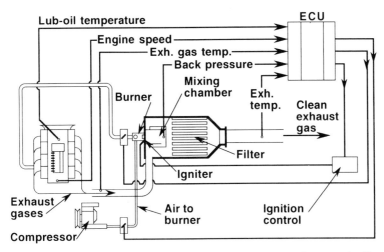

Fig. 8.11 The Deutz DPFS filter system has a diesel fired burner operated at intervals of between 10 and 13 hours while the vehicle is in running

by the filter rises above a predetermined level, the burner ignites. The temperature of the gas flowing through is limited by automatic regulation of the burner. Even so, it rises to as high as 600–650 deg C.

Deutz also use on their trucks a particle filter system developed by Waschkuttis GmbH, of Wiesenthau. This has one or more ceramic monolith cores. Regeneration can be effected either by a diesel burner or by means of an iron-based fuel additive similar to that used by VW, but introduced either during refuelling or fed continuously by a metering system into the delivery line to the injection pump.

Similar equipment has been developed by Zeuna-Stärker, of Augsburg, though with uncoated core elements. This, however, is available only with a diesel burner operating at 550°C, and combustion is supported entirely by residual oxygen in the exhaust gas. Electronic control is effected on the basis of signals of back-pressure, gas temperature ahead of the filter, and engine speed and load. Switching the burner on at intervals of about every 100 km for periods of about 3 minutes, it uses 0.5 litres of fuel per burn. It is claimed to reduce particulates by up to 90%, dependent on the physical dimensions and porosity of the core selected. This type of unit has also been used in a system comprising a pair of parallel filters with a single burner.

Typical of the sequentially operated dual filter systems is that of Voest-Alpine, of Linz, in Austria, Fig. 8.12. Its electronic automatic control, integrated into an engine management system, switches the exhaust gas flow alternately from one filter to the other. Consequently, regeneration can be effected while the vehicle is running. Exhaust back-pressure and temperature, and engine speed and coolant temperature are the inputs for the electronic control and monitoring functions. A blower, not shown in the illustration, supplies extra air for regeneration. There are four throttle valves, two to

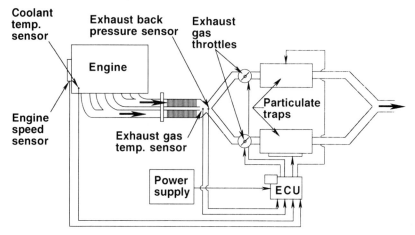

Fig. 8.12 Voest Alpine sequentially operated dual filter particle traps. Flexible tubing connects each section of the two-piece exhaust manifolds to the system

divert the flow of exhaust gas sequentially from one filter to the other, and one in each of the ducts that supply the two filters with air from the blower. The filters are uncoated ceramic monoliths and heat for regeneration is provided by electric elements. This system is claimed to have a filtration efficiency of 86%.

Iveco too has a twin sequential system, though it can also be used with a single filter. The filters are uncoated ceramic monoliths, but a diesel fuelled burner and a blower provides a copious supply of heated air. A filtration efficiency of 90% is claimed, and the maximum operating temperature is 900 deg C.

The MAN sequential system, Fig. 8.13, developed in conjunction with Leistritz AG, of Nuremberg, has twin uncoated ceramic multi-tube element monolith filters. A single diesel fuelled burner is used and, for regeneration, its output is diverted by the electronically controlled, pneumatically actuated valve that also diverts the exhaust gas flow from one filter to the other. Combustion is supported by air from an independent blower, and the temperature of regeneration is in excess of 550 deg C. Following regeneration of one filter, the hot air is diverted initially to the other before the burner is switched off and, subsequently, either one or, if needed, both filters are available for use until the next regeneration cycle.

Webasto, of Stockdorf, have developed a modular electronically controlled burner system, which they call the Soot Converter, for application to filters of either ceramic honeycomb, or wound ceramic yarn, or expanded ceramic core type elements. It comprises a diesel fired burner, upstream of the filter, a combined fuel supply pump and ignition initiator set, and a small engine-driven compressor, all controlled by a microprocessor based monitoring and diagnostic unit. The burner produces temperatures up to 700 deg C ahead of the filter. Combustion is supported by the excess air in

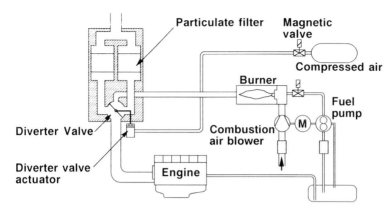

Fig. 8.13 The Mann system, developed in co-operation with with Leistritz AG, has twin uncoated monolith tubular element filters

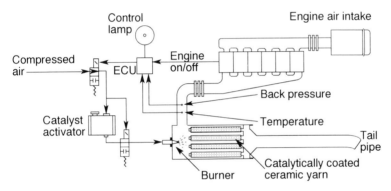

Fig. 8.14 In the Mann and Hummel system, developed in co-operation with Mercedes-Benz has stainless steel filter elements wound with copper-coated ceramic yarn

the exhaust gas, supplemented when necessary by air from the compressor. A typical regeneration cycle takes about 5 minutes, during which the fuel consumption is about 0.33 litres.

Mann & Hummel's system, developed at Ludwigsburg, in co-operation with Mercedes-Benz, functions on an entirely different principle. Its filter is of copper coated ceramic yarn wound round a set of perforated stainless steel tubes, Fig. 8.14, and it is regenerated chemically. The copper coating acts as a catalyst, reducing the temperature at which the particulates will burn off to not higher than 250 deg C. Burn-off is initiated by injecting periodically into the filter measured quantities of acetyl acetone, and combustion is supported by the excess air in the exhaust gas. A maximum of 95% and an average of 80% reduction in particulate emissions is claimed.

Ernst Aperatebau, of Hagen, similarly have tubular filter elements assembled parallel to each other in a heat and corrosion resistant casing, but the tubes are of porous ceramic and regeneration is effected by an electronically controlled, diesel fired burner. During regeneration, extra air is supplied from the engine-driven compressor. A filtration efficiency of 95% is claimed.

8.8 Influence of fuel quality

In Sweden, the aromatic hydrocarbon content of the fuel commercially available in 1989 was about 35–40% and the sulphur content 0.2%. However, Volvo was running its own factory service vehicles in and around the Gothenburg area on a special high quality fuel containing only 0.002% sulphur and 5% aromatics. Its cost was 20 p per gal more than the conventional fuel available locally, but NO_x emissions were more than halved and HC, CO and particulates reduced by at least 70–80%.

How individual emissions are influenced by different fuel properties have been summarised by the UK Petroleum Industry Association as follows:

NO_x Increases slightly with cetane number
      Decreases as aromatic content is lowered
CO   No significant effects
HC   Decreases slightly as cetane number increases
      Decreases with density
      Relationship with volatility inconsistent
Black smoke Increases with fuel density and decreases with aromatic content
      Is not significantly affected by volatility
      Increases with injection retard (e.g. for reducing NO_x)
Particulates Reduced as volatility is lowered
      Reduced as cetane number is lowered, though inconsistently
      Unaffected by aromatic content
      Reduced as sulphur content is lowered.

A good quality fuel is generally regarded as one having a cetane number of 50 and a sulphur content of no greater than 0.05%. Volvo has shown that bringing the sulphur content down from 0.2% to 0.05% reduces particulate emissions by up to 20%. Furthermore, in 1989, such a reduction would have cost only about 2 p per gal, before tax, whereas to obtain a commensurate improvement by reducing the aromatic content and increasing the cetane number would cost about 22 p per gal. On the other hand, the former measure would also reduce NO_x. The influence of fuel quality on the formation of NO_x has already been discussed in Section 8.2. Its effect on other emissions will be dealt with in the following Sections.

8.9 Black smoke

In the introduction to this Chapter and Section 8.5 are some comments upon the effect of sulphur content on the formation of particulates. As regards visibility, however, the carbon content is much more significant. Other factors include volatility and cetane number. In fact, claims that volatility *per se* influences black smoke are, as briefly mentioned in Section 8.2, without foundation. Smoke is reduced with increasing volatility for two reasons: the first is the correspondingly falling viscosity, and the second the associated rising API gravity of the fuel. A consequence of the first is increased leakage of fuel through the clearances around both the pumping elements and the injector needles and, of the second, the weight of the fuel injected falls. Therefore, for any given fuel pump delivery setting, the power output decreases with increasing volatility. In fact, the real influence of volatility depends on an extremely complex combination of circumstances, and varies with factors such as speed, load and type of engine.

The reason is that each engine is designed to operate at maximum efficiency over a given range of speeds and loads with a given grade of fuel. Therefore, at any given speed and load, a change of fuel might increase the combustion efficiency, yet at another speed and load the same change might reduce it. This is because a certain weight of fuel is required to produce a given engine power output so, if the API gravity is increased, the volume injected must be increased, and this entails injection for a longer period which, for any given engine operating condition could have either a beneficial or detrimental effect on combustion efficiency. Similarly, the resultant change in droplet size and fuel penetration relative to the air swirl could have either a beneficial or detrimental effect on combustion efficiency.

The reason why cetane number does not have a significant effect on the output of black smoke is simple. It is that smoke density is largely determined during the burning of the last few drops of fuel to be injected into the combustion chamber.

8.10 White smoke

White smoke, as mentioned in Section 8.3, is a mixture of partially vaporised droplets of water and fuel, the former being products of combustion and the latter arising because the temperature of droplets fails to rise to that needed for ignition. It can be measured by passing the exhaust through a box, one side of which is transparent and the other painted matt black. A beam of light is directed through the transparent wall on to the matt black surface. If there is no white smoke, no light is reflected back to a sensor alongside the light source, and the degree of reflection therefore is a function of the density of the white smoke. The instrument can be calibrated by inserting a standard matt white card into the box to cover the matt black surface, and the reading of an ammeter measuring the current from the sensor is taken to be 100%. When the card is removed, the ammeter reading is taken to be zero. For testing fuels, the criterion is the time taken, after starting from a specified low temperature, for the smoke level to reduce to an acceptable level.

Doing some of the early work with this type of meter, described in 'Diesel Fuel Properties and Exhaust Gas — Distant Relations', McConnell and

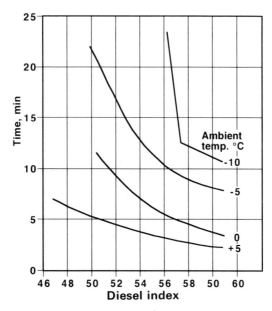

Fig. 8.15 BP have produced these curves showing the influence of Diesel Index on the time required for white smoke density to drop to an acceptable level after starting the engine from different temperatures

Howells, 1967, BP took acceptable to mean a drop to a 5 milliamp reading in five minutes. Illustrated in Figs. 8.15 and 8.16, are respectively the effects of Diesel Index and Cetane No. on white smoke, as obtained by BP. After starting at 0 deg C, satisfactory smoke levels are generally obtainable with a Diesel Index of 57 and a Cetane No. of 53.5.

8.11 Emissions in practice

Scania has carried out a great deal of research into emissions in practice. This has been reported in their excellent publication entitled 'The Diesel Engine and the Environment', published in 1991, which embraces an extremely wide range of factors influencing emissions output. Emissions vary greatly with load, so the results obtained from the test cycles laid down by legislation are suitable only for comparing the exhaust cleanliness of one engine installation with others.

Among the factors that affect emissions are the nature of the road surface, aerodynamic efficiency of the vehicle, topography of the route, type of operation, and driving technique. For example, emissions on logging operation partly on good metalled roads and partly on soft forest tracks can be as much as 30–40% higher than operation solely on well maintained highways. Again, NO_x emissions from a 25 tonne vehicle operating in top

EXHAUST EMISSIONS

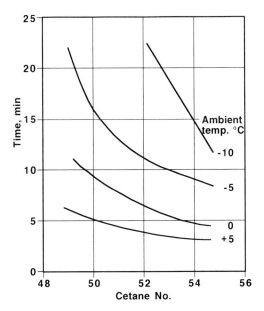

Fig. 8.16 Influence of Cetane No. on the time required for smoke density to drop to an acceptable level (410 words)Captions for Chapter 9

Table 8.2 NO_x EMISSIONS ON LONG DISTANCE TRUCK OPERATION, ASSUMING ENGINE EFFICIENCY $= 0.40$ AND $q_{NO_x} = 8.5$ g/kWh

Vehicle type	Gvw tonnes	Fuel cons. g/kWh	Fuel cons. g/km	NO_x emission g/tkm	NO_x emission g/t km	NO_x index g/t km
6-wheel rigid	30	17	2.9	9.9	0.58	100
6-wheel rigid + 4-wheel trailer	40	25	3.3	11.2	0.45	77
6-wheel rigid + 6-wheel trailer	52	34	3.9	13.3	0.39	67
6-wheel rigid + 8-wheel trailer	60	41	4.3	14.6	0.35	60

gear at 70 km/h on the highway has been shown to be 10 gm/km, while that from the same vehicle on stop–go operation in heavy traffic in an urban situation was 27 gm/km. This demonstrates the desirability of constructing urban bypasses and facilitating urban traffic flow.

From statistics drawn for a wide range of trucks on an equally wide range of types of operation, Scania have established the formula below, which they say predicts with adequate accuracy the quantity of NO_x emitted from a diesel powered truck. Using this formula they have drawn up in Table 8.2, which demonstrates the advantage, as regards reduction of NO_x, of operation with heavy vehicles.

$$Q_{NO_x} = E \times h \times q_{NO_x}$$

where Q_{NO_x} is the quantity of NO_x emitted, in g/km
E is the energy consumption, in kWh/kW, obtained by assuming an actual fuel consumption of 10 litres/100 km is the equivalent of 1 kWh
h, the mean efficiency during normal driving, is assumed to be 0.4%
and q_{NO_x} is the specific NO_x emissions of the vehicle in g/kWh.

Chapter 9

Test cycles, sampling, and analysis of exhaust emissions

Emissions regulations, testing, sampling and analysis in relation to gasoline engines has been covered in Vol. 1. Consequently, there is no need to include here much on the subject of CO, HC and NO_x. However, there is a great deal more to be added about international regulations relative to diesel powered vehicles and on the sampling, measurement and analysis of soot, particulates and sulphur.

9.1 Legislative control of diesel engine emissions worldwide

Although international harmonisation of emissions regulations is so clearly desirable, attainment is difficult because not only does the quality of fuels available vary around the world but so also do engine design philosophies, traffic conditions, speed limits, and driving techniques. This is especially true of diesel powered vehicles. For example, as was mentioned in the context of combustion chamber design in Section 7.8, whereas the European driver keeps close to the engine speed range giving maximum torque, his American counterpart tends to keep his foot down and let the governor determine the speed.

National and regional test cycles, therefore, have been developed to represent as closely as possible local conditions: compare for example the flat plains of Holland, the mountains of Switzerland and the vast distances and huge urban conurbations of the USA. Thorough analyses of and comparisons between the European, US and Japanese test methods have been published in a review by Cartellieri, Kriegler and Schweinzer, in Proc. I.Mech.E, Part D3, 1992, Vol. 206, p161.

9.2 Units of measurement

For the measurement of emissions, various units have been used. They have included parts per million (ppm) and grams per test. Now, however, the units of measurement of diesel engine emissions for trucks differ from those for diesel and gasoline powered cars. For the cars and light commercial vehicles, they are measured in g/km, but for the heavier vehicles, powered by engines developing more than 150 kW, the units are g/kWh. The latter units reflect more realistically the wider range of power outputs called for in the operation of the heavier vehicles, especially in congested traffic.

9.3 Historical review

Although the UK was the first country to set legal limits on smoke from exhausts, this early legislation was largely to mollify public opinion regarding its nuisance. Otherwise, as was the case with emission control of spark ignition engines, the USA, and in particular California, has led the way. Even so, the USA did not have any legal limits on smoke and particulates for cars and light trucks until 1986 and for heavier vehicles, over 3850 kg, until 1987.

European emissions limits for cars and minibuses have generally come into force for all new type approvals on October 1 of the year quoted. There were none for heavy vehicles until 1992, but for light vehicles, the 1982 regulations according to EEC R 15/04 are presented in Table 9.1. They were complicated by the fact that the limits increased with weight of vehicle, however no limits were included on smoke and particulates. In 1990, provision was made for tightening the limits for all such vehicles of less then 1.4 litres capacity as follows: to 45 gm/test for CO, 6 for NO_x and 15 for $CO + NO_x$ and, in 1991, to 30 g/test for CO and 8 for $HC + NO_x$. In October 1989, a limit of 1.1 g/test for particulates was introduced for all cars and DI diesel engines larger than 2 litres capacity and under 2.5 tonnes gross weight. Proposals for 1994 are: CO 2.72 g/km, $HC + NO_x$ 0.97, and particulates 0.14 g/km.

In October 1989, what are known as the 5th Amendment Directives EEC 87/76 and 88/436 came into force for cars powered by diesel engines of more than 2 litres capacity. By the end of 1992, the Consolidated Emissions

Table 9.1 THE 1982 EMISSIONS LIMITS IN G/TEST, SET OUT IN EEC REGULATION R 15/04, FOR DIESEL POWERED CARS AND BUSES CARRYING LESS THAN 6 PEOPLE

Weight, kg	CO	$HC + NO_x$
Up to 1020	58	19.0
1020–1250	67	20.5
1250–1470	76	22.0
1470–1700	84	23.5
1700–1950	93	25.0
1950–2150	101	26.5
Over 2150	110	28.0

Note: for buses carrying more than 6 people and trucks carrying more than 3.5 tonnes but weighing from 1020 to over 2150, the limits for $HC + NO_x$ are respectively 23.7, 25.6, 27.5, 29.3, 31.2, 33.1 and 35.0 g/test

Table 9.2 TYPE APPROVAL EMISSIONS STANDARDS, GM/KM, FOR LIGHT DUTY DIESEL POWERED CARS

Authority	Year	CO	$HC+NO_x$	Partics.
EEC Directive 91/441	1992	2.72	0.97	0.14
DI derogation of 40% on all pollutants until 1994				
MVEG* proposal	1996-7	1.0	0.7	0.08
DI derogation for 3 years as follows:		1.0	0.9	0.1

*Motor Vehicle Emission Working Group

Table 9.3 EUROPEAN EMISSIONS LIMITS TO 1992 AND PROPOSED TO 1999, FOR VEHICLES POWERED BY ENGINES OF MORE THAN 150 kW OUTPUT

Directive	Year	NO_x	HC	CO	Partics.
88/87	1990	14.4	2.4	11.2	—
EC Stage 1	1992	8.0	1.1	4.5	0.36
EC Stage 2	1995	7.0	1.1	4.0	0.15
EC Stage 3	1999?	5.0–5.5	1.1	4.0	0.10–0.15

Table 9.4 US PROPOSED EMISSIONS LIMITS FOR LIGHT VEHICLES, G/KM

Year	Non-methane HC	CO	NO_x	Partics.
1994	0.16 (0.19)	2.1 (2.6)	0.62 (0.78)	0.05 (0.06)
2004	(0.08)	(1.1)	(0.12)	(0.05)

Note: Figures in brackets are for a durability of 160 000 km, all others for 80 000 km

Directive EEC 91/441 became effective. These limits, and the recommendations for subsequent reductions, are shown in Table 9.2.

For vehicles having engines developing more than 150 kW, EC Directive 88/87 came into effect in 1990, and this has been followed by further reductions, as shown in Table 9.3, limiting CO to 11.2, HC to 2.4, and NO_x to 14.4 g/kWh for all commercial vehicles regardless of engine weight. Limits for the USA are set out in Tables 9.4, 9.5 and 9.6 and those for Japan in Table 9.7. Note that, for the reasons already given, the units used for specifying emissions for cars and trucks differ; furthermore, the earliest regulations specified emissions limits in terms of parts per million (ppm).

Table 9.5 US PROPOSED TYPE APPROVAL EMISSIONS LIMITS FOR LIGHT COMMERCIAL VEHICLES, G/KM

Year	Non-methane HC	CO	NO_x	Partics.
Up to 1700 kg weight				
1994	0.16 (0.19)	2.1 (2.6)	0.62 (0.78)	0.16
1995	0.16 (0.19)	2.1 (2.6)	0.62 (0.78)	0.05 (0.06)
2004	(0.08)	(1.1)	(0.12)	(0.05)
From 1701 to 2608 kg				
1994	0.20 (0.25)	2.7 (3.4)	(0.60)	0.08
1995	0.20 (0.25)	2.7 (3.4)	(0.60)	0.05 (0.06)

Note: Figures in brackets are for a durability of 160 000 km, all others for 80 000 km

Table 9.6 US FEDERAL LEGISLATION FOR TRUCKS, G/KWH

Year	HC	NO_x	Partics.
1980	1.74	14.34	0.804
1990	1.74	8.04	0.804
1991	1.74	6.7	0.335
1994	1.74	6.7	0.134

Notes: The particulates limits for city buses became 1.3 g/kWh in 1983.
Sulphur content of the fuel was taken as 0.1% for 11991 and 0.05% for 1994.
The US Senate proposed a limit of 5.4 g/kWh for NO_x by 1998.

Table 9.7 JAPANESE EMISSIONS LIMITS, MAX/MEAN, FOR DIESEL CARS AND TRUCKS

Year	Wt, kg	HC	CO	NO_x	Partics.	Smoke
Cars and minibuses carrying up to 10 passengers, g/km						
1992	<1250	0.62/0.4	2.7/2.1	0.98/0.7	—	50%
	>1250	0.62/0.4	2.7/2.1	1.26/0.9	—	50%
	Reducing by 1994 to respectively			0.72/0.5	0.2	40%
				0.84/06	0.2	40%
Vehicles carrying more than 2.5 tonnes, ppm						
1983		670/510	980/790	610/470	—	50%
1984		for indirect injection engines 390/290			—	50%
Vehicles carrying more than 2.5 tonnes, g/kWh						
1994		—	—	6.0/6.0	0.7	40%
		For indirect injection		5.0/5.0	0.7	40%
Proposed		—	—	4.5/4.5	0.25	25%

Note: Smoke is tested in three stages during acceleration at full load

9.4 Test cycles

Some of the test cycles for Europe, the USA and Japan have been illustrated in Vol. 1, Chapter 14. Among these, however, the Japanese 10 mode cycle has since been changed and will be explained here. Perhaps it should be further pointed out that, whereas the US FTP tests are based on recordings of actual journeys considered to be typical of the worst of the majority of driving situations, the European and Japanese cycles, comprising sections of constant acceleration and deceleration and constant speed, are devised to represent normal situations in road traffic. All are most conveniently run with computer controlled equipment, but it is possible, though not easy, to run the European and Japanese tests manually.

The new Japanese test is similar to the old, in which the 10 mode cycle was run through six times but measurements taken only in the last 5. Now, however, it is run three times and a higher speed stage has been substituted for the last three steps, as shown in Fig. 9.1. This part of the cycle is not unlike the European hot transient cycle, but rises to twin peaks of about 70 km/h. To avoid confusion it should perhaps be added that whereas the full cycle is shown in Fig. 9.1, only one element of it was illustrated in Vol. 1, Chapter 14. The 11 mode cold start cycle still has to be run through 4 times. Incidentally, for Europeans, neither translating nor interpreting the Japanese regulations is easy, and errors are therefore difficult to avoid.

Several test cycles have been introduced for heavy commercial vehicles. In the USA, the Constant Volume Sampling (CVS) method of the FTP 72 test was applicable to light commercial vehicles as well as cars. For diesel powered vehicles, however, the diluted gas had to be heated to 190 deg C to avoid condensation of the higher boiling point hydrocarbons as well as of water. As was the case for the gasoline engine tests, to avoid the condensation of water vapour in the gas, it remains necessary also to dilute

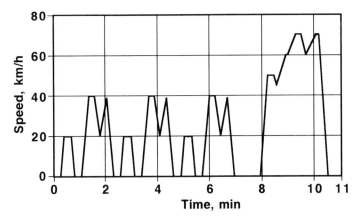

Fig. 9.1 The new Japanese 10 mode cycle comprises the first three parts of that which it replaces, followed by a 15 mode high speed, or hot, test. Total time 660 s, distance 4.16 km, idling condition 31.4% of total time

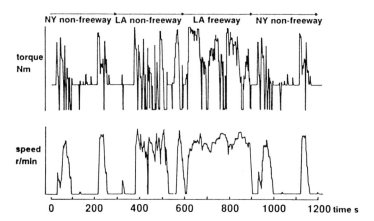

Fig. 9.2 In the US, the FTP 75 test is applicable for diesel powered commercial vehicles over 3.8 tonnes gvw (3856 kg). It is based on recordings of the actual road conditions illustrated here

the exhaust gas prior to collecting the samples for separation and quantitative analysis. Because of the dilution, highly sensitive measuring equipment must be used. The FTP 75, 13 mode test supplanted the FTP 72 test and, in 1984, the CVS Transient Test Cycle was introduced, but for diesel powered heavy commercial vehicles over 3.8 tonnes (8500 lb). This was based on recordings of actual road conditions, with both engine speed and torque varying continuously with time, Fig. 9.2. From 1984 to 1986, either engine or chassis dynamometer testing were permissible, but in 1986, engine dynamometer tests became mandatory for commercial vehicles. The sampling equipment had to be capable of carrying considerably higher throughputs of diluted gas than that used for cars.

An outcome of the introduction of particulate limits for the 1986 Transient Test was that a turbulent dilution tunnel had to be introduced into the sampling equipment, to prevent deposition of the solid content before it was passed through the filter. The turbulence is maintained by arranging for the Reynolds Number R (which is the density x velocity/viscosity of the fluid) of the tunnel to be greater than 40,000 up to the point at which the particulates are actually deposited on the filter paper.

In Japan, a 6-step test is in force for vehicles weighing more than 2500 kg (Fig. 9.3). In both instances, the mean value of the results obtained in the individual steps is recorded, and quoted in relation to the average engine power over the whole of the test cycle.

In Europe, ECE regulation R 49 applies to the testing of heavy duty commercial vehicles for gaseous emissions, Fig. 9.4. It is similar to the American 13 step test, but the weighting factors for the various stages differ, to make it representative of the operating conditions in Europe, mainly as regards frequency distribution of load and speed.

At the time of writing, only the USA has regulations regarding particulates, but in all the major countries, separate test cycles have been introduced for measuring black smoke. In most instances, the assessment has to be made

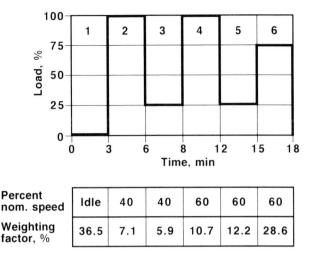

Fig. 9.3 The Japanese 6-step test for vehicles weighing over 2500 kg

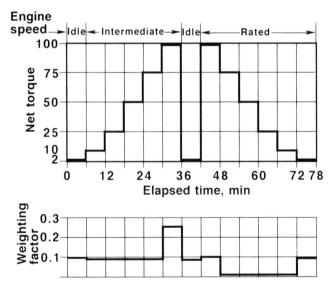

Fig. 9.4 ECE regulation R49, for the testing of heavy commercial vehicles for gaseous emissions is similar to the US 13-step test, but the weighting factors are different

on an engine test bench. The USA Federal Smoke Test cycle is illustrated in Fig. 9.5. Its six stages are run through three times, the arithmetic mean of the smoke values being recorded during only the acceleration stages, 2, 4 and 6. Step 2 is maximum acceleration against only the inertia of the engine and unloaded dynamometer, up to about 90% of nominal speed. In step 3, the dynamometer load is increased to bring the speed down to about 60%

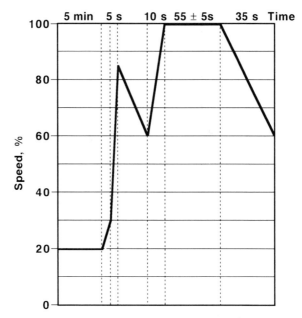

Fig. 9.5 The US Federal six mode smoke test cycle. The third (a 10 sec) stage is done with the accelerator pedal released

of nominal speed. Next, step 4 is maximum acceleration against full load, from 60% to 100% of nominal speed within a given time. Step 6 is deceleration at full load, within a given time, effected by increasing the braking load to cause the speed to fall from 100% to the higher of either 60% of nominal speed or the speed at which maximum torque is developed. The smoke values must be measured by an opacimeter equivalent to that used by the US Public Health Service.

In Japan, all diesel powered vehicles (cars and commercial) must pass a common smoke test. The smoke level is determined at full load under equilibrium conditions at three speeds, Fig. 9.6. Measurements may be taken using either an opacimeter or a filter.

In Europe we have EC Smoke Test R24, the driving schedule is illustrated in Fig. 9.7. As with the other smoke tests mentioned, the condition of the engine is specified in great detail. For example, it must be fully run in, and equipped on the test bed with the exhaust system to which it will be connected on the vehicle. Moreover, the oil and water temperatures must be maintained at the levels at which it will normally operate. The engine is run at full load at six equal speed intervals from 45%, but not less than 1000 rev/min, to 100% nominal speed. Measurements have to be taken with an opacimeter having a specified absorption limit relative to nominal gas throughput. However, since it has its disadvantages, as explained in Section 9.7, the filter method of measurement is also, at the time of writing, under consideration for approval as an alternative.

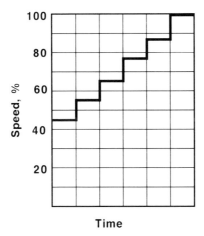

Fig. 9.6 For the Japanese smoke tests, all stages are at full load and they are terminated when their measurements have been completed. If the acceleration to 0.4 times maximum speed, again if stage 1, takes less than 16.66 sec, the stages must be completed in 16.66 sec

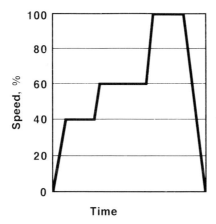

Fig. 9.7 The test cycle for the EEC R24 smoke test. All stages are at full load and are terminated when their smoke measurements have been completed, unless the acceleration to stage 1, at 0.45 maximum speed, is attained in less than 16.66 sec. In the latter eventuality, the stages must be completed in 16.66 sec

A full load acceleration test is also included in EC Regulation 24. This is to establish a figure for comparison with results that might be obtained in subsequent monitoring in the field. The engine is run up to normal operating temperatures and then accelerated over a period of between 2 and 5 sec from idling up to maximum speed, during which the only load is that imposed by its rotating masses.

9.5 Sampling the exhaust gases

The method of sampling and measuring the exhaust emissions from diesel engines is in general similar to that described in Volume 1, for gasoline engines, but there are some significant detail differences. Fig. 9.8 is virtually identical to Fig. 14.1 in Volume 1, though the gas collection bags are represented differently and the equipment added specifically for testing diesel exhaust emissions is enclosed by the dotted line. In fact, Fig. 9.8 is a diagrammatic representation of the whole of the installation designed for testing both gasoline and diesel engine emissions in the Shoreham laboratories of Ricardo Consulting Engineers Ltd. The layout, Fig. 9.9, showing the measuring section of the equipment was not shown in Volume 1, but is identical for both gasoline and diesel engines.

To ensure that none of the heavier fractions present in diesel fuel condense out before the HC content can be measured, the sampling probe S_2 in Fig. 9.8 is heated to a temperature of 190 ± 10 deg C throughout its length from the pick-up point to the flame ionisation detector (FID). Between the pick-up and FID, the sample is passed through a filter, similarly heated. This is to remove the particulates that might influence the results.

Smoke and particulate emissions are not a problem with gasoline engines For diesel engine testing, a pick-up, S_4, is required for sampling the particulate content. Again, the measuring equipment must be as close as practicable to the sample pick-up point but, in this case, so that the heavier particles do not fall out before they can be measured.

The measuring set-up in Fig. 9.9 is housed in a separate cell. Gas from the sample bags in Fig. 9.8 is drawn, by the pump P, through the filter F to the measuring equipment. Illustrated diagrammatically on the right, from top to bottom, in this illustration are the instruments for measuring HC, high CO for gasoline and low CO for diesel engines, CO_2 and NO_x. Of these, the HC measurement equipment, enclosed by a double chain dotted line, is not used for diesel engines because, as indicated in the second paragraph of this Section, a hot sample is required so the measurement is undertaken closer to the sample pick-up point. Details of the instruments are given in Sections 9.6 to 9.10.

From the pump P, the sample gas is delivered to a pipe that distributes it to the flow control valves N that serve each measuring instrument. From each of these valves, it is delivered to a three-way valve by means of which either it or the zero, or span, gases can be selected for passing through the measuring instruments. Nitrogen is generally used as the zero gas for all except for the HC instrument, for which air free from contamination with HC is suitable.

The other span gases are mixed with air in the proportions needed for setting the maximum values of pollutant to be recorded, and stored at high pressures in bottles. Dilution is necessary because, if each instrument had to be capable of recording the output from the largest engine likely to be tested, its range of measurement (span) would be so wide that obtaining readings of adequate accuracy for the small engines would be impossible.

ANALYSIS FOR DIESEL EXHAUST EMISSIONS

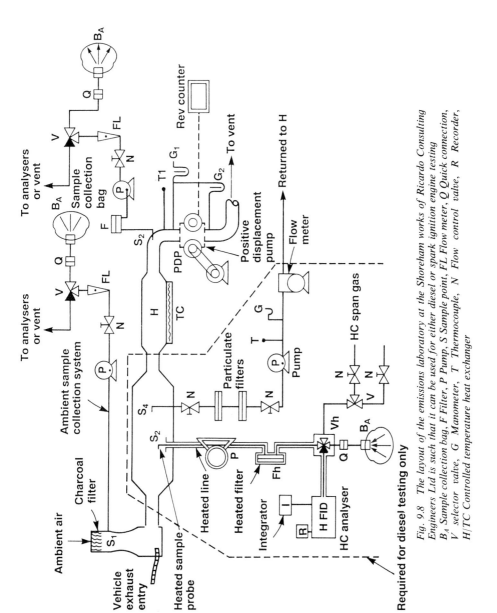

Fig. 9.8 The layout of the emissions laboratory at the Shoreham works of Ricardo Consulting Engineers Ltd is such that it can be used for either diesel or spark ignition engine testing B_A Sample collection bag, F Filter, P Pump, S Sample point, FL Flow meter, Q Quick connection, V selector valve, G Manometer, T Thermocouple, N Flow control valve, R Recorder, H/TC Controlled temperature heat exchanger

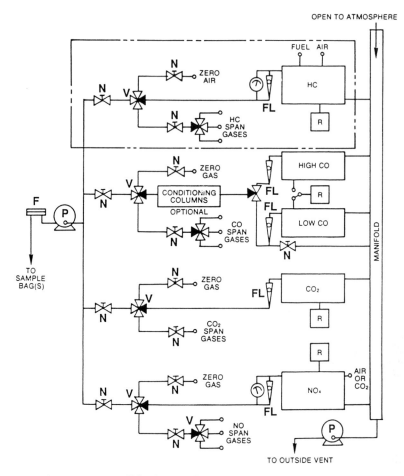

Fig. 9.9 The measurement cell for the Ricardo emissions testing laboratory. The key letters used are identical to those listed below the caption to Fig. 9.8

From the instruments, the sample gases are discharged into a manifold, in which they are mixed with air drawn in by the pump in the bottom right-hand corner of Fig. 9.9. Thus diluted, they are discharged though a vent pipe outside the building.

The particulates are weighed, the CO and CO_2 content assessed by non-dispersive infra-red (NDIR) absorption analysers or ultra-red absorption recorders (URAS), NO_x is measured by the chemiluminescence (CLD) method, and the hydrocarbon content measured with a flame ionisation detector, Fig. 9.10.

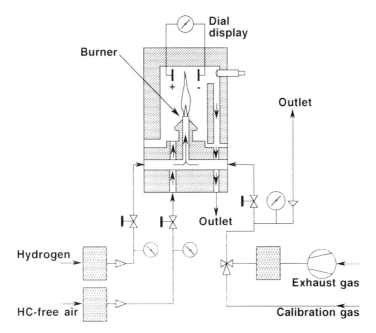

Fig. 9.10 Flame ionization detector. Because of the high boiling point fractions present in diesel fuel, the temperature of the sample gas must be maintained at 190 ± 10 deg C right up to the burner. Consequently, heat exchangers, as well as filters may be needed for conditioning the gases in the supply circuits

9.6 CO and CO_2

For meeting the increasingly stringent requirements for detecting the minute quantities of CO in diesel exhaust gas, the Signal Series 2000 non-dispersive infra-red (NDIR) analyser, Fig. 9.11, is more suitable for analysis than the ultra-red absorption recorder (URAS) equipment. It is based on the fact that each gaseous compound can be identified by the infra-red radiation waveband that it absorbs. Moreover, the energy of the absorbed radiation increases the temperature of the gas and therefore, if it is held at constant volume, also its pressure. In contrast, the elemental gases such as nitrogen and oxygen do not absorb the radiation.

In the instrument, two infra-red light beams of equal intensity are passed through a motor driven light chopper, similar in principle to that in the equipment described in Vol. 1, into parallel channels each comprising a number of components. First the light passes through a filter cell in which the irrelevant wavebands are removed. One of the two beams issuing from

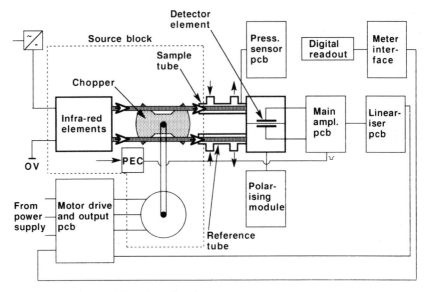

Fig. 9.11 Signal Series 2000 infra-red analyser.

the filter is passed through a cell containing the sample of the gas to be tested, while the second passes through the other cell, which contains nitrogen. That passing through the nitrogen does so without any significant loss of energy, but energy will have been absorbed by that going through the sample. Next the two infra-red beams enter the detector cell.

As can be seen from the illustration, the detector cell is divided, by a thin metal diaphragm, into two chambers both of which are filled with the gas to be detected (CO or CO_2). The beam that has passed though the nitrogen contains more energy than that which has come through the sample cell, so there will be a temperature, and therefore pressure, differential between the gases on the two sides of the diaphragm. A Luft, or microphone capacitance, detector senses the change in capacitance due to the deflection of the diaphragm, and converts the pressure difference across it into a voltage.

The AC output from the detector is amplified and rectified to convert it into a DC voltage. This voltage, which is proportional to the concentration of the gas being measured, is modified by a span potentiometer, and the combined signal is displayed digitally on an instrument calibrated to indicate CO concentration instead of volts. Since the level of absorbed energy depends upon the number of molecules present, and therefore the pressure in the cells, a sensor and amplifier are incorporated to compensate for day-to-day variations in atmospheric pressure.

The instrument is claimed to have a response of 90% full scale in less than 10 sec, with a linearity of $\pm 1\%$ FSD (full scale deflection) and a repeatability of better than 1% FSD. Its sensitivity is said to be better than 100 ppm FSD, with a resolution of 1 ppm. Optional extra equipment includes a sample pump, span/zero selection valves, auto zero, high/low alarms, range identification and remote range change. This equipment is also suitable for measuring concentrations of other gases such as CO_2, NO, NO_2 and SO_2.

9.7 HC and NO_x

Nitrogen oxides are measured by the chemilumiscence method, Fig. 9.12. When nitric oxide (NO) reacts with ozone, light is emitted in the waveband 590 to 3000 nm. This is the gas that constitutes the major proportion of the oxides of nitrogen in the exhaust gas, though it is accompanied by significant proportions of NO_2. Other oxides of nitrogen are present in little more than the proportions found in the ambient atmosphere.

Before measurement, the NO_2 and other oxides have to be converted to NO, so the exhaust gas is passed through a thermo-catalytic converter before being forwarded to the reaction chamber. In this chamber, ozone (O_3) is mixed with the NO to generate the chemiluminescence. To avoid false readings due to the luminescence of other chemicals in the gas, only radiation in the band 600 to 660 nm is measured. The signals from the sensor are amplified by a photoelectric multiplier, processed by an electronic system and the readings shown on a visual display.

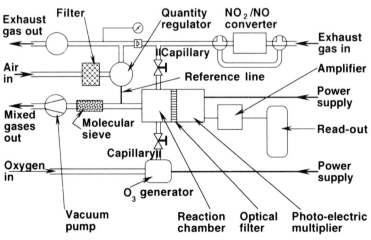

Fig. 9.12 Chemiluminescence equipment for measuring NO_x, based on the fact that when ozone reacts with NO, light between the wavelengths of 590 and 3000 nm is emitted; the filter blocks all except those between 600 and 660 nm

9.8 Measurement of particulates and black smoke

There are several methods of measuring black smoke. One entails passing a measured quantity of the exhaust gas through a filter paper, which is either weighed before and after (the *gravimetric* method), or the light reflected from it is measured by a reflectometer, to indicate the density of contamination. Weighing has the disadvantage that it does not distinguish between the carbon and other constituents of the particulates, including oil mist, most of which have only a marginal effect on the visibility of the smoke. Also, filter papers are hygroscopic, so measurements have to be taken under controlled conditions of humidity and temperature. The adsorption of water by glass fibres in filter paper, however, can be reduced by treating them with teflon. Despite these drawbacks, the gravimetric method is specified in the USA for their particulate test. Another method of assessment is comparison of the colour of the deposit on the filter paper with those on a standard grey scale.

More commonly, an opacimeter is used to measure the absorption of light passing through the sample of exhaust gas. Opacimeter readings are affected by the distribution of the soot particles and their grain size. Moreover, the frequency response of the sensor to the light varies with the voltage of the electricity source, so this must be strictly controlled. These disadvantages, it would appear, are no more severe than those of using filter papers so, as mentioned previously, the opacimeter is specified for the European smoke tests.

9.9 The Hartridge smoke meter

The Hartridge Smokemeter Mk 3, Fig. 9.13, is an instrument in which a beam of light is passed through the exhaust gas flowing past an opacimeter and recordings taken of how much light is absorbed absorbed by it. Provided there is neither oil mist nor water vapour in the exhaust gas, the readings increase logarithmically with the soot concentration The Hartridge instrument is the only opacimeter mentioned in BS Au 141a(1971). It essentially comprises two optically identical tubes mounted parallel, one above the other: one, through which clean air is passed, serves as the datum, while the exhaust gas sample is passed through the other. The bores of both tubes are non-reflective. A beam of light is directed axially through each tube in turn on to a photoelectric cell, the difference in the two light readings being a measure of the density of the smoke. Both light source and photoelectric cell are mounted on a frame that can be rotated to transfer simultaneously the beam and sensor from one tube to the other.

The output from the photoelectric cell is amplified and then measured by an ammeter the scales of which are calibrated to read from 0 to 100 Hartridge units and 0 to infinity absolute units of light absorption. Zero is set by adjusting a potentiometer in the meter circuit.

A small fan blows air into the clean air tube, whence it passes out over

ANALYSIS FOR DIESEL EXHAUST EMISSIONS

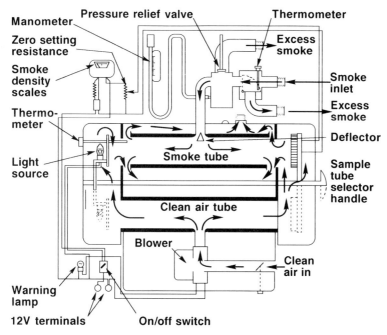

Fig. 9.13 In the Hartridge smoke meter the light source and photoelectric cell are mounted, at opposite ends, on a rotatable frame, so that they can be positioned axially in line with the gas and air tubes alternately. The two tubes, which are non-reflective, can be removed for cleaning

the light source and photoelectric cell, to prevent them from being fouled by the exhaust gas. It is then delivered into the compartment housing the sample tube and mixes with the exhaust gases that issue from its ends, before passing out through a pipe to atmosphere.

Either plain or electrically heated sample tubes can be installed in the machine, electric heating being necessary to comply with EEC Directive 72/306. This equipment has been certified as complying with ISO 3173, EEC Directive 72/306, Annex 7 and ECE regulation, Annex 8.

9.10 AVL 415 smoke meter

Before describing the instruments for measuring smoke content on the basis of filter paper blackening, some basic principles need to be understood. First, it is primarily the graphitic carbon, or soot, content that is measured by the reflectometer, and this content is a function of the blackening of the filter paper. Again, the measured values increase logarithmically with the soot concentration. The reflectometer reading from the clean white paper is zero and that from completely blackened paper, 10. If P_b is the degree of blackening of the paper, R_w the reflective power of the white paper and R_b that of the blackened paper, then:

$$P_b = (1 - R_b/R_w)$$

For the determination of the *Filter Smoke Number* (*FSN*), the volume of gas sampled must be taken into account. Since this is proportional to the length of the column of exhaust gas drawn through the filter paper, we need to know only the effective length of that column (L_{eff}) at some specified pressure and temperature. The FSN smoke level units are similar to those formerly used by Bosch, but are calculated according to ISO 10054 Draft Proposal, which specifies a length of 405 mm at 1 bar and 25 deg C. However, for other column lengths, an expression for the equivalent concentration, in mg/m^3, of soot in the gas can be calculated from the degree of paper blackening by applying conversion factors available in a Table. In this Table, the measurements are based on data established in the UK by the Motor Industry Research Association and confirmed by both the University of Darmstadt and AVL List GmbH.

The AVL 415 Smoke meter, Fig. 9.14, is a modern example of the filter paper type in which measurement is made by a *reflectometer*. Before each sample is measured, the system is purged by opening solenoid valves to direct clean air through it in a direction opposite to that of the exhaust gas flow during measuring operations. Then the solenoid valves reverse the direction of flow so that a diaphragm pump can draw the exhaust gas in through a probe inserted into the exhaust pipe and take it on through a filter paper in the metering head. This paper is delivered from a roll, first to the metering head and, subsequently, to the reflectometer head. Signals of sample temperature and pressure are passed from sensors to an electronic control, which regulates the diaphragm pump to deliver to the paper precisely the required quantity of exhaust gas.

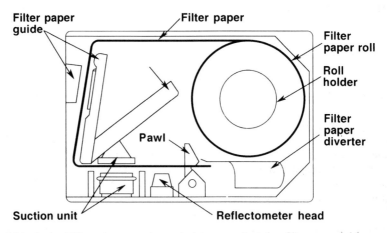

Fig. 9.14 In the AVL smoke meter, the particulates are collected on filter paper, fed from a roll. The paper is first clamped in the suction unit while the sample is drawn through it and then passed over a reflectometer head

ANALYSIS FOR DIESEL EXHAUST EMISSIONS

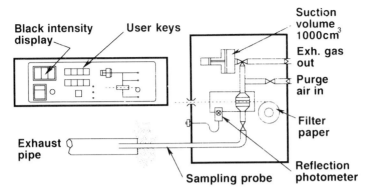

Fig. 9.15 The Pierburg PR1 smokemeter sucks a sample of 1000cm³ from the exhaust tailpipe and passes it through a filter paper, which is then moved on to a reflection photometer for assessment. Left: electronic control cabinet; right, sampling box

Signals from the reflectometer are transmitted to the electronic control unit, which converts them into Filter Smoke Numbers and displays them digitally. The measuring range is 0–10 FSN, and the readings are accurate to 0.01 FSN and with a reproducibility of 0.05 FSN.

An advantage of this diaphragm pump arrangement in association with an electronic control is that the quantity of gas sampled can be changed according to the density of the smoke. By switching automatically from one quantity to another, it is possible for the meter to be operating always in its optimum measuring range. Since the rate of sampling is virtually constant, the measured values represent the mean value of the soot content of the exhaust gas over that period of time. A feature of this instrument is that it can be powered from the mains, a built in rechargeable battery, or the battery of the car being tested, so that it can be used on the road or on a test bed.

9.11 Pierburg smoke meter

The Pierburg PR-1 smoke measuring device, Fig. 9.15, is another good example of the type in which the light reflected from the contaminated paper is measured photoelectrically. A 1000 cm³ sample of exhaust gas is passed through a filter paper, and the grey value of the deposit on the paper assessed by an integrated electronic system. To correspond with the Bosch blackening scale for a sample of 300 cm³, the total reflection from the 26 mm diameter uncontaminated surface exposed to the exhaust gas is assigned the value of 0 and, for zero reflection from the same surface when totally blackened, the value of 10. The whole scale is then linearised electronically to suit the 1000 cm³ sample taken by the equipment. Special filter paper is supplied in rolls and, each time a new test is started, it is automatically fed into position for exposure to the exhaust gas.

When the instrument is first switched on, it automatically performs a function check to ensure that the required suction is available, the paper is in position, the photometer is operating properly, and that the ambient conditions are appropriate. Then, before each measurement is taken, the whole system is purged and the instrument recalibrated. The resolution obtainable is either 0.1 or 0.01 units, and the precision $\pm 1\%$ of the measuring range. Deviations of measurements from tolerance limits are displayed electronically. This instrument can be used on the road as well as on an engine test bench.

Chapter 10

Optimising air induction — variable valve timing and differences in approach for diesel and gasoline engines

As has been explained in Chapter 7, the diesel engine is at a disadvantage relative to a gasoline engine because its fuel has to evaporate and be thoroughly mixed with the air in the extremely short time between the start of injection and the piston's coming up to top dead centre. On the other hand, by virtue of compression ignition and an essentially greater degree of swirl and macro-turbulence, combustion is initiated at a number of points in the combustion chamber. Consequently, the generation of heat more uniformly throughout the charge expedites evaporation and, because the multi-point initiation of combustion generates micro-turbulence, mixing is further promoted.

Even so, as the driver's demand for torque rises to its peak, i.e. increasing proportions of fuel are injected, it becomes progressively more difficult to burn completely all of the charge. Indeed, it is impracticable to burn more than about 80% of the air that can be made available. Consequently, the only way of enabling this type of engine to compete with its spark ignition equivalent is to supply extra air.

There are several ways of optimising air throughput. Most costly of all, though the most effective, is to supercharge or turbocharge, as explained in Chapter 12. The least costly, and only marginally effective, is to design the induction manifolding in a manner such that it offers as little resistance to flow as possible. This aspect has been adequately dealt with in Vol. 1. A more effective low cost method is to utilise the pulsations and standing waves in the manifold to augment the pressure difference across the valve port, thus inducing more air to flow through it than otherwise would do so. This is dealt with in Chapter 11. Yet another, which is the subject of this Chapter, is to incorporate a variable valve timing mechanism to optimise the quantity of air inducted relative to the fuel input, but this is considerably more costly than the alternatives.

All the measures outlined above, with the exception, in some respects, of turbocharging, have been developed to more advanced stages for gasoline than for diesel engines, which is why the mechanisms and principles are discussed mainly in relation to the gasoline engine. The reason why the application to the gasoline engine is so far ahead is that hitherto it has been subjected to more stringent exhaust emission and fuel economy regulations. Now, however, the attention of the legislators has turned towards the diesel engine.

Variable valve timing has the following additional attractions for application to naturally aspirated diesel engines. First, the timing can be optimised to increase volumetric efficiency and therefore power and torque. Secondly, the valve overlap period can be reduced during low speed operation, to increase effective expansion ratio, and improve idling stability and cold starting, as well as to reduce emissions throughout the low speed light load range. Advantage of variable valve timing in relation to turbocharging are outlined in Section 12.9.

10.1 The Atkinson Cycle

The current edition of Chambers Science and Technology Dictionary defines the Atkinson cycle as 'one in which the expansion ratio exceeds the compression ratio; more efficient than the Otto cycle but mechanically impracticable.' Since that edition was published, however, variable valve timing mechanisms have become available or in production, so it can now be said to be practicable.

Ideally, the Atkinson cycle entails full expansion of the charge. In diesel engines, however, as both load and peak combustion pressure decrease, over-expansion can occur following heat loss to coolant during the compression and power strokes. Under light load, therefore, partial throttling, to reduce the density of the charge, might be desirable. Overall efficiency could be further improved by introducing variable compression ratio, though problems relating to cost and potential unreliability mitigate against its application to high speed engines.

10.2 History of variable valve timing

The concept of variable valve timing (VVT) is far from new. Since 1880, almost 800 patents on this subject have been issued in the USA alone. In an SAE Technical Paper No. 890764 by Dresner and Barkan, the systems proposed have been classified into 15 basic types. Elsewhere, they have been further categorised also under three main headings: variable phase control (VPC), combined valve lift and phase control (VLPC), and variable event timing (VET) systems. At the beginning of the 1990s, interest was revived as a result of legislative pressures to reduce further the undesirable constituents of exhaust gases, including CO_2. Reduction of the latter of course implies substantial enhancement of thermal efficiency and thus reduction in specific fuel consumption.

Until recently, variable valve timing has not been commercially successful, because its complexity and cost has been unacceptable. Additionally, most of the mechanisms proposed exhibit other severe disadvantages such as, major increases in the dynamic loading of the valve train, rapid wear, noise, limited effectiveness, and large frictional losses. However, some commercially viable systems are now available, notably that of Honda.

10.3 Two types of variable valve timing

Although variation of inlet valve closure has been investigated, variable valve timing is generally applied in one of two ways: either the point of inlet valve closure is fixed and that of its opening varied, or both are fixed relative to each other but their timing (i.e. the inlet phase) advanced or retarded simultaneously, generally by rotating the cam relative to the shaft. The latter, termed phase change, has two advantages. First, the dynamic loading of the valve gear is unchanged and, secondly, the mechanisms for varying the timing are generally considerably less complex.

If the opening only is varied, retarding its timing not only reduces the overlap but also reduces the open period. This shortens the time available for opening the valve. Consequently, unless the lift is reduced, it inevitably increases the acceleration loading on the valve gear, which can cause problems at high speeds.

APPLICATION TO GASOLINE ENGINES

10.4 Early or late inlet valve closure

With fixed valve timing, designers mostly aim at high nominal output, with the result that because the inlet valve is still open after bdc, Fig. 10.1, torque at the lower end of the speed range is impaired owing to back-flow losses.

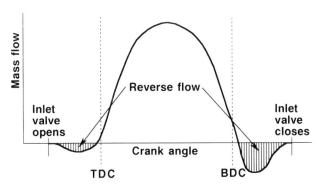

Fig. 10.1 The significance of back-flow into the induction manifold in an engine running at low speed and with fixed valve timing can be seen in this diagram of mass flow of air plotted against crank angle

In contrast, Mercedes-Benz, for example, have found that, by optimising inlet valve closure over the whole of the speed range, improvements in full load torque of up to 14% and maximum torque by 4% can be obtained. However, if the torque is maximised at maximum power with wide open throttle, the initial period of rising torque, or back-up, as the throttle opening is reduced is impaired or lost.

At the 1992 FISITA Congress in London, paper C389/188 presented by Crouch et al, of Ricardo Consulting Engineers, explored the potential for the application of variable valve timing from two distinctly different viewpoints: the enhancement of the thermal efficiency of throttle controlled engines; and as a means of obviating throttling losses. The following are some of the findings.

Running a typical modern 2 litre engine under part load conditions, reductions of approximately 8% in bsfc, 25% in NO_x and 30% in HC emissions were obtained by closing the inlet valve earlier than normal. At full load, an increase of about 2% in bmep was obtained by closing it 15 deg later than normal.

During low speed operation, both inlet valve closure (IVC) and exhaust valve opening (EVO) should be relatively close to bdc. Indeed, for both stability and economy at idling, the overlap should be almost zero. Under all other conditions, larger overlaps are needed to increase the breathing potential and reduce output of NO_x. Going further, by utilising the valve control system to eliminate the throttle as a means of regulating the torque output, overall fuel economy can be improved by 5 to 6%.

Pumping losses during idling represent 30–40% of the total so, while the throttle remains closed, their elimination could improve fuel consumption by up to 15%. Even by simply reducing valve overlap from 10 to zero deg for idling, an overall fuel economy of 2% in fuel consumption can be obtained. As the load is increased and the throttle opened wider, the pumping losses fall and the benefit of course becomes less. However, since a high proportion of motoring is done at light load, the attraction of abandoning the throttle control for gasoline engines is obvious.

10.5 Late inlet valve closure (LIVC)

In another paper, No. 389/041, presented at the 1992 FISITA Congress, O'Flynn and Saunders of the University of Sheffield and Ma of the Ford Motor Company discussed late inlet valve closure (LIVC) as a means of controlling torque output. With late closure, part of the charge is returned into the induction manifold as the piston passes bdc. In these circumstances, the gas exchange occurs at approximately constant pressure, so only the proportion of charge trapped in the cylinder after inlet valve closure is compressed. As the throttle is opened, the valve closure is progressively brought forward to the standard timing for the engine.

For the tests conducted by the Authors, a 4-cylinder gasoline engine with a 4-valve head and multi-point injection was used. Late closure, however, was effected on only one of the two inlet valves. Four extra cams on the exhaust cam shaft actuated the inlet valves at fixed timing, while four on the inlet cam shaft actuated the late closing valves. All the cams had identical profiles. Phase change was introduced by altering angle of the timing sprocket relative to that of the crankshaft but, to avoid backfire and contact between the inlet valves and pistons, its range was limited to 140 deg.

If the inlet valve closure is very late, the quality of combustion can be inferior. One reason is that, because of late in-port mixing, multi-point fuel injection is at a disadvantage relative to carburation. This disadvantage can be overcome if the volumes of the primary induction pipes (branch pipes) are large enough to contain the return flow into the manifold so that there is no possibility of their being robbed, by an adjacent induction pipe, of some of the mixture they contain following injection into them. Alternatively, the required effect can be obtained by limiting the lateness of inlet valve closure.

The Authors found that, with multi-point injection, the volume displaced by the piston before inlet valve closure should not exceed about 80% of that in the primary pipe and port. On the engine used for these tests, it was found that with conventional throttle control, No. 1 cylinder became weak, Nos. 2 and 3 rich and No. 4 remained constant as the load was reduced, Figs. 10.2 and 3. On the other hand, when LIVC control was instituted, not only was the degree of maldistribution much greater but also of an entirely different pattern: Nos. 3 and 4 went lean while 1 and 2 became rich. Consequently, to avoid misfire, especially in No. 1 cylinder, increased fuel consumption was unavoidable.

At light load, the maldistribution is exacerbated because the minute slug of fuel injected per cylinder is likely to be both inadequately mixed with air and, with currently conventional fuel injection, inaccurately metered. The back-flow into the plenum can sometimes contain most of the fuel that has been injected and at others virtually none, so running becomes unstable.

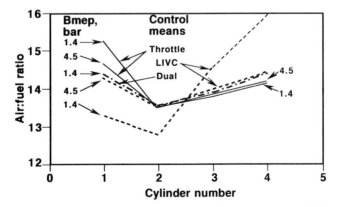

Fig. 10.2 Comparison of air:fuel ratio variations between the four cylinders at 3000 rev/min with throttle, LIVC and dual control

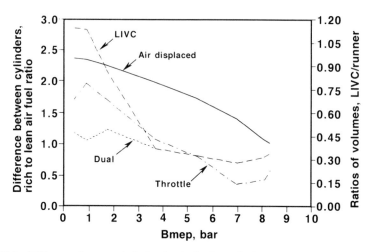

Fig. 10.3 Differences between air:fuel ratios in the four cylinders at 2000 rev/min

Operating with LIVC from medium to full load, flame initiation (over 1% mass fraction burnt) is more rapid than with conventional throttling control, but the rate of flame spread from 1% to 90% burnt is the same. Also, the spark needs to be advanced about 4 deg less than with conventional throttled operation and, by virtue of lowered compression ratio, knock sensitivity is reduced. At loads lower than 1 bar bmep, the spark occurs before IVC; indeed, at 3000 rev/min, it was as much as 17 deg before.

The best reduction in fuel consumption obtained with LIVC on this engine was 5.7% at 1000 rev/min and 2.73 bar bmep, Fig. 10.4. At lower loads, maldistribution and other effects inhibited further improvements. However, using both LIVC and throttle control, the latter to reduce maldistribution and increase the rate of vaporisation, the best reduction was 8.0% at 2000 rev/min and 0.94 bar bmep. The lower combustion pressures and temperatures experienced with LIVC were accompanied by reduced NO_x. On the other hand, HC output was considerably increased with this engine, even with dual control (LIVC and throttling), though no explanation could be found.

10.6 Early inlet valve closure (EIVC)

In FISITA paper C389/333, Geringer et al, of Mercedes-Benz, and Ebersbach et al, of Daimler-Benz, pointed out that varying the valve timing and lift, again of a gasoline engine, could increase the full throttle torque by up to 14%. Furthermore, by substituting early inlet valve closure for throttling, part load fuel consumption in steady state operation can be reduced by as much as 9–15%. The potential for reduction in pumping losses under idling conditions is illustrated in Fig. 10.5. For maximum overall benefit, the inlet

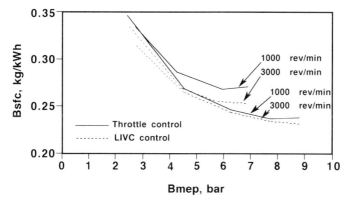

Fig. 10.4 Comparison of rates of fuel consumption with LIVC and throttle control at 1000 and 3000 rev/min

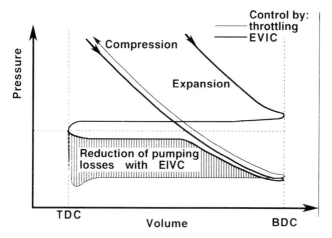

Fig. 10.5 On this part of an indicator diagram for an idling engine, the shaded area represents the pumping loss reduction due to LIVC

timing has to be continuously variable down to closure throughout the cycle. However, a degree of accuracy, or reproducibility of control, approaching $\pm 1°$ crankshaft angle is desirable, and this is extremely difficult to attain.

For the engine tests, variation of the timing of IVO was effected by rotating the camshaft relative to the crankshaft but, for early closure, hydraulic tappets were used, in which the hydraulic pressure could be dropped at will, Fig. 10.6. Such a dual control system has the advantage that valve lift is initiated by the ramp on the cam profile and the degree of lift can be reduced by varying the rate of leakage during the valve open period. However, if flow restriction is used for controlling the rate of leakage from the tappet during closure, gentle seating of the valve is not easy to maintain throughout the life of the engine. In practice, therefore, some noise might be difficult to avoid

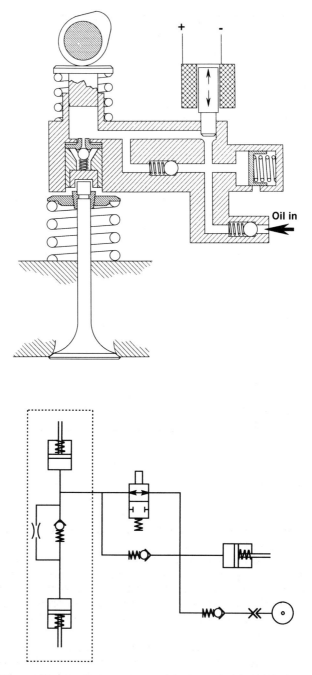

Fig. 10.6 The variable valve timing system used in the work done jointly by engineers from Daimler-Benz and its subsidiary Mercedes-Benz

OPTIMISING AIR INDUCTION — VARIABLE VALVE TIMING

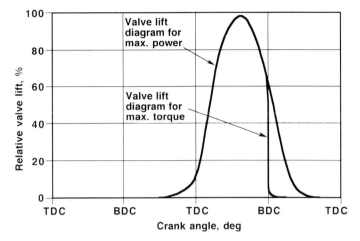

Fig. 10.7 At full load, a broad valve-lift curve is needed for obtaining good nominal power output, but rapid closure at about bdc is better for maximum torque

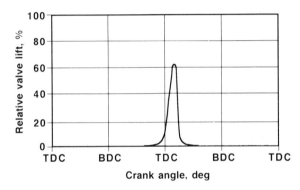

Fig. 10.8 Inlet valve lift at part load with control by early inlet valve closure instead of a throttle valve

especially as, owing to temperature variations, the viscosity of the oil varies with temperature and other factors.

With high lift and rapid valve closure, pumping losses are at a minimum and charging efficiency is at its optimum. From Fig. 10.7, it can be seen that, for good nominal output, the valve-lift curve is broad but, for maximum torque, the lift is high and the valve closes rapidly at approximately bdc. At light loads without dual control, the valve-lift curves are of much smaller area, as in Fig. 10.8.

By shutting off one valve in a multi-valve head, swirl can be enhanced and charge quality improved. Further gains can be obtained by incorporating variable induction pipe tuning, stratifying the charge, and cylinder shut-down by, for example, keeping both the inlet and exhaust valves shut. As regards enhancement of overall performance, however, relatively little is to be gained from varying the exhaust valve lift and timing.

Because variable inlet valve timing can reduce final compression and therefore combustion temperatures, it offers potential for reduction in NO_x. Moreover, by virtue of the higher induction manifold pressure, mainly during idling and slow running under light load, the pressure gradient between the inlet and exhaust ports is higher, so the residual gas content of the combustion chamber is lower, and idling stability better.

10.7 Problems associated with EIVC

Although early closure of the inlet valve, especially when used as an alternative to throttle control of gasoline engines, Section 10.8, increases overall efficiency, is not without its problems. First, for gasoline engines, high pressure in the inlet port is less favourable for mixture formation than are the low pressures associated with throttle control. Thus the films of fuel deposited on the walls of the ports and manifold are thicker, so large drops of fuel are drawn intermittently into the combustion chamber, some settling on its walls. The outcome is wide variations in air:fuel ratio, and therefore low efficiency and high emissions.

Very early closure of the inlet valve limits the flow into the combustion chamber to the period during which the piston speed is low, and therefore adversely affects both the flow pattern and turbulence within the chamber. Subsequent expansion further slows down the swirl and mixing. As the piston moves down to bdc, expansion can cool the gases to as low as -15 deg C, causing recondensation, which can continue into the early part of the compression stroke. Indeed, the total drop in temperature due to condensation can be as high as 50 deg C. Owing to incomplete and intermittent mixing, the condensation starts abruptly at different points in the combustion chamber. It increases with decreasing load but, as the speed falls, leaving more time for heat transfer from the walls of the combustion chamber, it decreases.

10.8 Variable valve timing with control by throttle only

In general, fixed valve timing is optimised either for the speed at which maximum torque is developed or to obtain good nominal power output. For controlling a gasoline engine by varying the valve timing, on the other hand, the valves are opened for periods that decrease as speed and load are reduced. Three benefits are thus obtained: the incoming gas cannot blow back into the induction system, pollution of the charge by residual exhaust gas is reduced, and optimum use is made of the energy liberated by the combustion of the charge.

As the load is increased, the increasing duration of opening improves the volumetric efficiency. With increasing speed, the valve open periods are

OPTIMISING AIR INDUCTION — VARIABLE VALVE TIMING

extended, ultimately providing the gas throughput needed for maximum power at full throttle. Additionally because, with varying engine speeds, times available for gas interchange expressed in terms of crank angle differ from those in terms of time, it is desirable to vary the inlet and exhaust event phasing independently relative to crank angle.

VVT MECHANISMS

10.9 Some simple systems

Variable phase control can be effected in a number of ways. One is to advance and retard the cam shaft by means of a sliding muff coupling on a divided shaft, with spiral splines on the driven and straight splines on the drive interfaces, or vice versa. This, however, suffers the disadvantage of frictional resistance to control operation. Another method is to install in the belt or chain drive to the camshaft a movable idler pulley in combination with a tensioner having a longer than usual stroke. Movement of the idler pulley towards or away from the drive, Fig. 10.9, rolls it around the half-speed wheel to advance or retard the timing while, at the same time, the tensioner compensates for the movement. In general, this appears to be the most practicable and least costly variable valve timing system.

Some of the least complicated alternatives include, for instance, devices reducing the lift of the inlet valve under light load and low speed conditions of operation. This not only limits the quantity of gas entering the cylinder but also obviates the losses associated with the throttling the otherwise

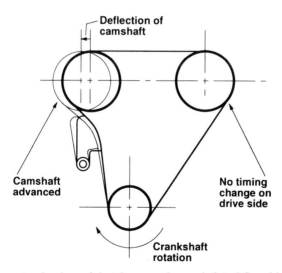

Fig. 10.9 For varying the phase of the inlet cams, the camshaft is deflected horizontally, thus causing the half-speed wheel to roll relative to the belt

potentially large inflows of air. Moreover, it could also increase the turbulence and swirl of the charge in the cylinders. Valve lift can be varied by increasing and decreasing the distance between the cam shaft and tappets or rocker ends within the limits dictated by the need for the valve to close completely. Decreasing the lift in this way, however, causes the follower to clear the ramp of the cam during opening and closing, with consequent increases in wear and noise.

Increasing and decreasing the lift is generally done by varying the length of hydraulic tappets, as on some Honda 1983 motor cycle engines. Incorporation of a leak orifice can automatically decrease the lift as speed is reduced. Problems also arise not only as regards noise, as previously explained, but also owing to variation of viscosity of the oil with temperature.

Another simple means of controlling valve movements is by use of high power solenoids, such as Colenoids or Helenoids. With these devices too, however, it is difficult to avoid impacts between valves and seats. Moreover, their efficiency is low, so energy losses are severe: peak powers of about 10 kW per valve are needed and, averaged over the two revolutions of the Otto cycle, the energy consumed amounts to perhaps 400 W per valve.

10.10 VPC, VLPC and VET systems

Variable phase control implies simultaneously advancing and retarding the inlet valve opening and closing points through equal angles while retaining those of the exhaust valve as they were, Fig. 10.10. This varies the overlap so that low speed torque and, with it, specific fuel consumption are improved over most of the speed range. Since the duration of opening remains constant, wide open throttle power is unaffected.

A combination of lift and timing control (VLTC), Fig. 10.11, can offer further performance enhancement. On the other hand, it is more costly than VPC. One way of combining these two is to have axially stepped cams, and effecting the variation by shifting the followers from step-to-step, but the mechanism is complex. Tapered cams, such as the Fiat Tittolo system combining variable lift and event timing, are an alternative but this means virtually point contact between cam and follower and, if the duration of opening is kept constant, they are extremely difficult to manufacture. Moreover, the axial loading introduced is about 10% of the force applied by the cam to the follower, and the point of contact between cam and follower has to be shifted axially by the controller while subjected to this axial loading.

Honda have a commendably simple system in production, Fig. 10.12. It has three cams and rockers per pair of valves in a four-valve head but, at low speeds, only the outer pair of rockers actuate the valves, leaving the central one idling freely. At high speeds, the electronic control signals a hydraulic valve to open, to allow oil pressure to move a plunger which locks all three rockers together. In this condition, since the central cam lobe is bigger than the two outer ones, the latter idle and the former actuates the valves.

OPTIMISING AIR INDUCTION—VARIABLE VALVE TIMING

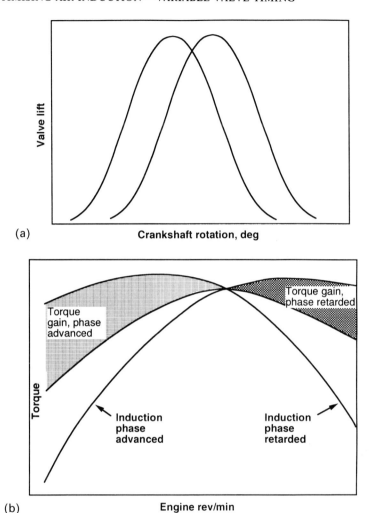

Fig. 10.10 With the variable phase control system (VPC), represented at (a) the inlet opening point can be retarded to give less overlap, for stability and low emissions at idle. For good low speed torque and lower fuel consumption, the inlet opening should be advanced, to increase the overlap. The differences in the torque characteristics are illustrated at (b)

Varying the valve event timing, VET, is the changing of the duration of lift while keeping the timing and magnitude of maximum lift constant, Fig. 10.13. In other words, only the opening and closing points are varied. This improves part load emissions and economy, leaving the wide open throttle condition unchanged, and has the advantage that the ramps on the cam remain effective both as the valve begins to lift and when it reseats. Again,

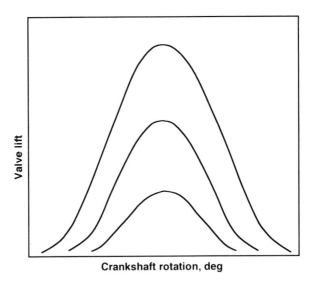

Fig. 10.11 With a stepped cam for providing both variable lift and timing (VLTC), extra advantages are obtainable. At the lower lifts, friction is reduced, better charge swirl achieved and fuel consumption improved. The illustration shows three steps, though up to ten are practicable. This system is suitable for application to rocker mechanisms

it is possible, as in for example the Mechadyne–Mitchell system, to combine VET with VPC.

10.11 Mechadyne–Mitchell system

The Mechadyne–Mitchell principle, developed by Mechadyne Ltd in close association with Kolbenschmidt AG, is applicable to almost any of the commonly used valve actuation mechanisms. It would appear to be the simplest system that does not have serious drawbacks, yet offers both variable event timing and phasing. Although it has a number of additional parts, almost all are identical for each cylinder, and therefore the extra tooling costs are not unreasonably high.

Basically, a hollow camshaft is driven by a peg on the outer end of a lever projecting from a drive shaft carried coaxially within it. This peg projects into a slot in the camshaft, Fig. 10.14. The drive shaft can be located axially by any of the devices used for locating conventional camshafts.

The arrangement is shown in greater detail in section BB of Fig. 10.15, from which it can be seen that the peg is actually a ball and slider reciprocating in a slot in a collar on the camshaft. The drive shaft can be moved laterally to vary the drive from concentric to eccentric. When driven concentrically, as at (a) in Fig. 10.14, the speed of rotation of the hollow shaft, and therefore the cams, is constant provided engine speed remains constant. However, when it is moved, say 5 mm off centre, as at (b), its instantaneous speed of

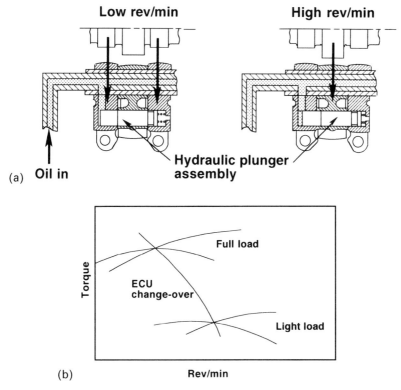

Fig. 10.12 Honda variable valve timing system. In the section on the left, the two outer rocker arms are actuated by their cams, while the central one is idling. That on the right shows the condition when electronic control unit (ECU) has signalled a solenoid to open a valve allowing oil under pressure to push the three-piece hydraulic plunger to the right, locking all three together, so that the central cam, because it is higher than the other two, actuates the valves. In the graph below, the central curve shows how the ECU varies the change-over point with speed.

rotation is multiplied in the ratio $2.4:1.9 = 1.263:1$. Therefore, as the shaft rotates, the ratio reduces progressively first to 1:1 at 90 deg, and then on to the inverse of 2.4:1.9, at 180 deg, and finally back again to complete the 360 deg. As the control is moved to increase the eccentricity of the drive shaft, the duration of lift of the valve is progressively reduced because its opening is retarded and its closing advanced.

Obviously, this can happen only with appropriate phasing of the cam relative to the eccentric. This is rendered possible by dividing the hollow camshaft, along its length, into the same number of sections as there are cylinders, each section being carried in its own bearings and driven by a separate lever and peg projecting radially outwards from the one-piece drive shaft. Incidental advantages of dividing the hollow camshaft into short sections are that the functioning of the ramps on the cams is unaffected and

270   OPTIMISING AIR INDUCTION — VARIABLE VALVE TIMING

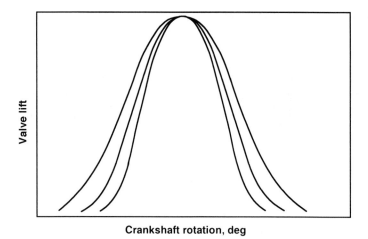

Fig. 10.13 This shows the characteristics of variable event (VET) timing, without phase change. Valve lift remains constant and the variation is continuous, as oppose to stepped. Part load emissions and economy are improved, while leaving unimpaired the wide open throttle condition. Characteristics of a VET system with phase change is illustrated in Fig. 10.17

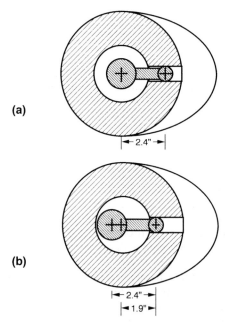

Fig. 10.14 Diagrammatic representation of the principle of the eccentric drive of the Mechadyne-Mitchell system, (a) in the coaxial and (b) eccentric position. In the latter position, rotation speeds up over about 25 deg each side of the nose of the cam, thus bringing closer together the valve opening and closing points, as in Fig. 10.11

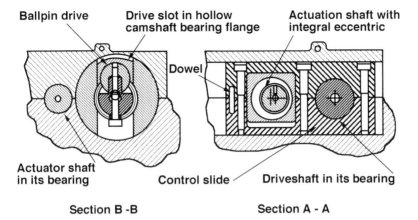

Fig. 10.15 Sections AA and BB taken from Fig. 10.16 to show in detail the ball-pin drive and actuation shaft

the short lengths of shaft are inherently very stiff both torsionally and in bending.

The mechanism by means of which the drive shaft is moved off centre is illustrated in the plan view of the 24-valve DOHC head of a six-cylinder engine, and Section BB in Figs. 10.15 and 16. Here, it can be seen that the solid drive shaft is driven in the conventional manner by a wheel mounted on the flange at the left hand end of the cylinder head. To one side of and parallel to the coaxial drive shaft and six-piece camshaft is the control shaft, which is actuated by a lever on its end projecting from the right hand end of the cylinder head assembly. Incidentally, this projection and the control lever mounted on it is the only part of the mechanism that cannot be housed within the standard valve cover. In production, the controller and actuator would almost certainly be incorporated in the cylinder head structure.

From Section BB it can be seen that a mechanism comprising an eccentric in a scotch yoke slides vertically in the two-piece casting that forms not only their housings but also the drive shaft bearings and caps. Rotation of the control shaft moves the whole assembly laterally and therefore the drive shaft into and out of concentricity with the cam shaft sections. The control shaft can be rotated through only 90 deg which is why, when it has been actuated to bring the drive shaft into its eccentric position, as in section BB, there is no clearance above the scotch yoke.

By arranging the belt or chain drive as in Fig. 10.9, lateral movement of the drive shaft can be made to cause also a phase change. The resultant changes of the valve timing are illustrated in Fig. 10.17, which shows what the lift characteristics are with the standard timing, what they would be if only event timing were changed and what it is when both event timing and phase shift are applied. The whole system can be applied to both the inlet and exhaust valves but, for optimum cost-effectiveness, Mechadyne suggest that only the inlet valve timing should be varied.

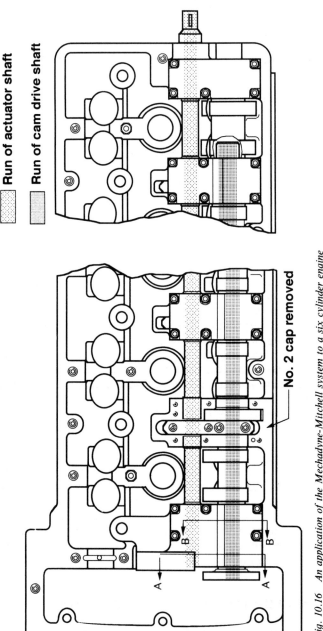

Fig. 10.16 An application of the Mechadyne-Mitchell system to a six cylinder engine

OPTIMISING AIR INDUCTION — VARIABLE VALVE TIMING

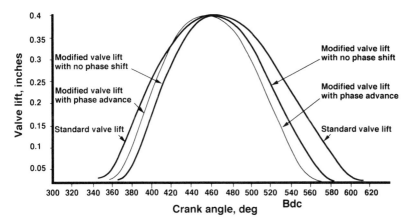

Fig. 10.17 Characteristics of the Mechadyne-Mitchell system with phase change

In an application to a Ford 8 valve DOHC engine, the eccentricity was 3.6 mm. The standard inlet cam characteristics were defined as 244 deg duration of lift, timed to start at 13 deg and terminate 51 deg btdc. Without any change in the timing of the maximum lift point (zero phase shift), the Mechadyne–Mitchell mechanism could have varied this timing from standard down to −10 and 28 deg btdc. On the other hand, with a cam designed for increased maximum power (lift duration 260 deg), a 13/67 timing could have been similarly varied down to −10 and 44 deg btdc. In both instances, no phase shift is assumed. In fact, phase shift was used and the actual ranges of timings chosen were −1 and 19 for the 244 deg cam and −1 and 35 for the 260 deg cam.

The range of valve opening per mm valve lift could be varied between about 230 deg for wide open throttle and 180 deg for low speed operation. For most automotive engines, these figures would probably translate into about 250 to 200 deg respectively.

The operating envelope is of course limited by the stresses superimposed on the valve train by the accelerations due to advancing and retarding the timing. However, since the eccentricity of the drive is introduced only at speeds of 2000 rev/min and below, the total stresses in the valve train are no higher than those with fixed valve timing at 4000 rev/min. Tests with twin cylinder motorcycles have shown 32% power increases and reductions in specific fuel consumption of 43%.

10.12 Control of the Mechadyne–Mitchell system

The friction to be overcome when moving the shaft eccentrically is not great, so electric or hydraulic power would appear to be the good alternatives. Electric power, however, though easily regulated by the engine ECU, is not necessarily the most acceptable since, even without it, there are already too many demands on electric systems.

Where hydraulic power is available, for example on power steered vehicles, this is more attractive. Alternatively, power might be taken from the engine lubrication system, though this would probably entail the introduction of a larger pump and a hydraulic accumulator, to avoid oil starvation in the engine bearings. Variations of viscosity with temperature would of course have to be taken into consideration and could present problems.

Low cost cars might have an on–off solenoid control, possibly based on only engine speed sensing. The more upmarket models and commercial vehicles, however, might have a more complex continuously variable eccentricity controller, with speed, load and VET position sensing for closed loop mixture control by the ECU.

Chapter 11
Optimising air induction —induction pipe tuning

Induction pipe tuning is among the measures to which more attention has been paid in connection with gasoline than diesel engines. Initially, this was because of some success with sports, racing and record breaking cars. More recently, however, it has been extremely widely adopted for family cars, partly because of the desire to improve fuel economy as well as performance. It is particularly applicable to diesel engines because of the previously emphasised need to optimise air induction. However, it has not yet been widely adopted for the latter, so all the practical examples given in Sections 12.11 to 12.16 are of gasoline engines. In Vol. 1 there was insufficient space for more than a brief mention of the principle, so it is dealt with in detail in this Chapter.

11.1 The three effects

Three entirely different principles are applied in induction pipe tuning. Of these, the easiest to visualise is the *inertia wave* effect, initiated by closure of the inlet valve. The incoming air piles up, under its own momentum, against the closed valve until the pressure generated causes it to rebound at sonic velocity as an acoustic wave. This is reflected back from the open end of the induction (or primary) pipe and, by the time it has returned, the inlet valve has opened again, so it flows on as a pressure pulse into the cylinder.

It is both longer and of larger amplitude than the higher frequency resonant (or standing) waves like those in an organ pipe, which constitute the second effect. These are initiated by the rapid opening of the inlet valves. If utilised at all, they are mostly applied by bouncing them back into the cylinder repeatedly while the valve is open, to modify the overall shape of the torque curve. Effective utilisation of both the inertia and standing waves entails using an induction pipe of a length such that the pressure pulses created by both arrive back while the inlet valve is open.

The third effect that has to be taken into consideration is inter-cylinder robbery of charge. Where two cylinders are inducting simultaneously, they are competing for air from the plenum chamber. Since all cylinders have to fire in two revolutions, the larger the number of cylinders the greater is the problem, and it can be the most significant of the three effects.

In short, induction system tuning is the combined art and science of optimising the effects of the inertia and acoustic waves in the induction pipes, by techniques that are perhaps more familiar in connection with the design of musical instruments. However, the possibility of both inter-wave and inter-flow interference between adjacent pipes has to be taken into account.

11.2 Resonant, or standing, waves

When the air in a pipe is transiently displaced axially, the wave that this displacement generates in it tends to bounce repeatedly from end-to-end, as indicated in the middle diagrams of Figs. 11.1, 11.2 and 11.3. Each of these diagrams shows the form of the resonant waves in a pipe of a different configuration. Their fundamental frequencies depend on the length of the pipe. Resonance can be initiated by, for example, an external force causing vibration of the pipe, or by some disturbance of the actual air in the pipe at any point along it or at its ends.

The principal relevant disturbance is caused by the opening of the inlet valve which, as it lifts off its seat, creates a suction pulse, possibly backed up by the depression in the cylinder. Because the air is elastic and has mass, it responds by surging forward to restore the pressure, thus initiating an alternating set of depression and compression pulses, which are in fact sound waves and travel along the pipe at the speed of sound. In print, this sequence

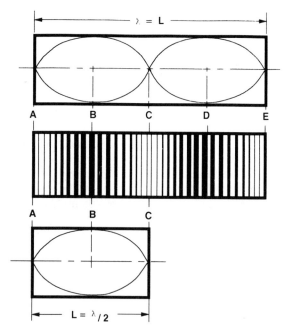

Fig. 11.1 Diagram showing the fundamental and first overtone mode of vibration of the air in a tube both ends of which are closed

OPTIMISING AIR INDUCTION — INDUCTION PIPE TUNING

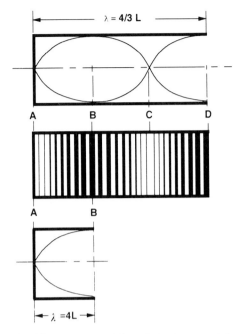

Fig. 11.2 Fundamental and first overtone mode in a tube having one end closed and the other open

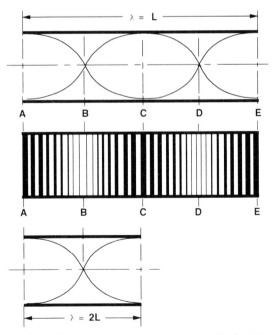

Fig. 11.3 Fundamental and first overtone mode in a tube both ends of which are open

of events could appear to be occupying a significant length of time; in reality, however, it is measured in millisec.

At low speeds and with valve overlap, there can also be a slight puff-back of gas from the combustion chamber into the inlet port, but this comes after the initiation of the sound wave, is less forceful and does not necessarily significantly interfere with it. Much depends on the degree of valve overlap. The larger the overlap, the longer is the period available for such pulses to have an effect. However, so far as induction system tuning is concerned, we are not concerned with low speed operation. During the valve overlap period, pulses from the exhaust system can have an effect too. The energy in the waves increases with speed of the engine, in particular because the lifting of the inlet valve off its seat is correspondingly more rapid.

In all circumstances, we have a resonant vibration phenomenon associated with a mass-spring system. The mass is that of the column of air in the pipe and the spring element the compressibility of that air. One wavelength λ is a complete cycle, or 2π radians, and therefore is equal to L in the top diagrams of Figs. 11.1 and 11.3, and $4L/3$ in Fig. 11.2. The phase difference of the displacement and pressure waves is always $\pi/4$, or 90 deg.

At this point, some clarification as to what exactly happens at the open ends of a pipe is necessary. On reaching the open end remote from the valve, a negative pressure wave sucks a slug of air in, while a positive pressure wave propels a slug out. In both instances these effects take place against the influence of atmospheric pressure and, owing to the inertia of the air flow, there is an over-swing and therefore a bounce-back. Both pressure and suction waves are reflected as pressure waves: in other words, the sign of a suction wave is inverted.

Provided the inlet valve is open by the time the waves reach the other end, both negative and positive pressure waves are discharged through the valve port into the cylinder. The depression in the cylinder is lower than that in the wave, and therefore tends to oppose bounce-back, though obstruction of the port by the valve head could have a minor influence.

If we plot the axial vibrations in the pipe to a scale such that the maximum amplitude of displacement in each direction equals the radius of the pipe section, they can be illustrated as shown in the top and bottom diagrams in these three illustrations. In each, the upper diagram represents the second order, and the lower one the first order, or fundamental mode of vibration.

From the upper diagram of Fig. 11.1, it can be seen that axial motion of the air is positively stopped by the closed ends, A and E, of the pipe. These ends are therefore displacement nodes. Mid-way between them is a third displacement node, while B and D are displacement anti-nodes. Because the air alternately moves towards and is bounced back from the displacement nodes, A and C and E are pressure anti-nodes. In other words, while the pressure remains constant at B and D, it fluctuates cyclically at A, C and E. This condition can occur in an induction pipe only when both a throttle and inlet valve are closed so, as regards manifold tuning, it is not of practical significance but it is relevant for automotive engineers concerned with body, cab or saloon noise.

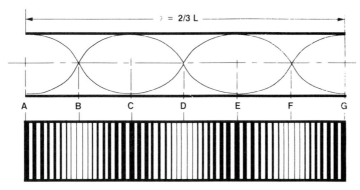

Fig. 11.4 The third harmonic in a pipe both ends of which are open

If one end of the pipe is open, Fig. 11.2, the air at that end is free to be displaced, so it becomes a displacement anti-node, which accounts for the different arrangement of the displacement curves for fundamental mode of vibration and overtones. This condition can arise when either the inlet valve is closed and the inlet pipe open, or *vice versa*. Therefore, this is a condition in which we are interested.

For a pipe open at both ends, the fundamental and first overtone harmonics are shown in Fig. 11.3. The third harmonic is illustrated in Fig. 11.4. Since this is a condition that arises only when both the inlet valve and pipe are open, it is of significance in relation to resonance effects initiated by the sudden opening of the inlet valve.

Clearly there must be some displacement beyond the open end before a reaction can occur, so a correction factor has to be applied to the length of the pipe. In fact, the effective length of an open end is L plus about 0.6 times its radius r so, for one open and one closed end, the correction factor is $L(1+0.6r)$ and, for a pipe with both ends open it is $L(1+1.2r)$. The time t taken for the completion of one wavelength is called the periodic time, or the period of the vibration, and the time required for the pulse to return to the inlet valve is $2L/c$, where c is the velocity of sound in the induction pipe. This, however, is rarely accurate in the context of induction system tuning, since the configurations of the ends of the passages are not those of a plain pipe ends, and there are some other influencing factors such as air temperature and diameter of pipe.

11.3 Pipe end-effects

Both the movement of the air into a pipe and its displacement due to the vibrations tend to cause turbulence around its open end, reducing the efficiency of flow. This can be avoided by flaring the end of the pipe to form a trumpet of approximately hyperbolic section, for guiding the air flow smoothly into it and thus increasing the coefficient of inflow. An increase in flow of up to about 2% can be thus obtained. The effective length of a pipe with such an end fitting is that of the parallel portion plus about 0.3–0.5 of the length of the flare. If the outer ends of the pipes terminate in apertures in a flat plate, or in the wall of a plenum chamber, their flares should extend well clear of the flat surfaces and also be clear of any adjacent walls. This is to ensure that the approach velocity is well below that within the pipe.

Another way of reducing this particular end-effect is to taper the pipe, increasing its diameter from the inlet port to its open end. This is sometimes done on very high speed engines, such as are installed in racing cars. The aim is at reducing the velocity of flow into the open end, and therefore the tendency for turbulence to be generated there. However, it may not be ideal for the generation of standing waves. Another important consideration, especially for high speed engines, is that reducing the velocity of flow also reduces the viscous drag between the air stream and the walls of the tube.

11.4 Frequencies, wavelengths and lengths of pipes

From the four illustrations, it is easy to see that the harmonic frequencies for pipes closed or open at both ends are $f_1, f_2, f_3, f_4, \ldots, f_n$, while those of the pipe closed at one end and open at the other are the odd numbers, $f_1, f_3, f_5, f_7, \ldots, f_n$. The formula from which these frequencies can be obtained is $f = c/\lambda$, where c is the velocity of sound in air and λ the wavelength. The frequencies of the first three modes of vibration in each case therefore are as follows:

Pipes with closed ends	$f_1 = c/2L$	$f_2 = c/L$	$f_3 = 3c/L$
One end open	$f_1 = c/4L$	$f_2 = 3c/4L$	$f_3 = 5c/4L$
Both ends open	$f_1 = c/2L$	$f_2 = c/L$	$f_3 = 3c/2L$

The velocity of sound in dry air is $\sqrt{\gamma p/\rho}$, where p is the gas pressure, ρ the density, and γ the ratio of the specific heats of the gas. At the standard temperature and pressure in free air, this velocity becomes 331.4 m/sec. Standard temperature and pressure is 298.15 deg K and 10^5 Pa (1 bar). Potential for some slight confusion arises, however, when referring back to data predating the universal introduction of SI units, at which point it became 273.15 deg K (0 deg C) and 101.325 Pa. At velocities of more than Mach 0.25, viscous friction losses impair engine performance.

Which ever version of the speed of sound in free air is taken, it is independent of frequency and, because pressure divided by density is constant, it can be considered also to be independent of variations of pressure, certainly of the magnitudes experienced in inlet manifolds. The velocity of sound varies with temperature according to the following relationship

$$c = c_0 \sqrt{(1 + \alpha \vartheta)},$$

where $c\vartheta$ and c_0 are the velocities of sound at ϑ and 0 deg C respectively, and α is the coefficient of expansion of the gas. In induction pipes, however, the velocity of sound is influenced by diameter, Fig. 11.5. Frequency is also

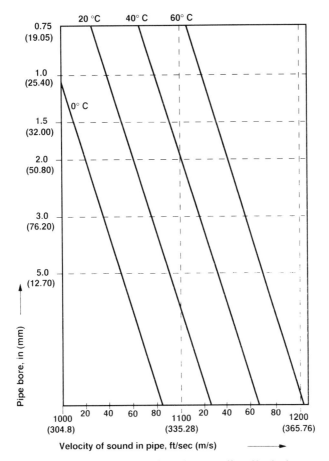

Fig. 11.5 Diagram showing how the velocity of sound in air is affected by the diameter of the pipe

affected, but relatively slightly, by the length:diameter ratio and internal smoothness of the pipe.

Since γ is dependent on the nature of the gas, the presence of fuel vapour, as in carburetted or throttle body injected spark ignition engines, will also affect the speed of sound in the manifold. Even so, this is not of practical significance, except possibly where the system comprises a set of straight tubes, since extreme accuracy is generally unattainable. Indeed, induction systems have to be optimised during development, for example by the use of telescopic elements.

Resonance increases the amplitudes of the pressure pulsations, though this effect is modified by damping. Damping can arise from roughness of the inner faces of the walls of the induction tract, the presence of bends, and obstructions such as throttle valves and inlet valve stems and guide ends. From damped and undamped resonance curves in Fig. 11.6, it can be seen

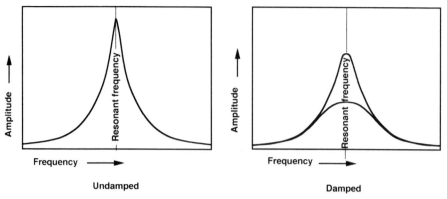

Fig. 11.6 Left, undamped resonance; right curves at two rates of damping

that the effect of damping is not only a reduction in maximum amplitude but also it rounds off the peak, and spreads the resonance over a range of frequencies that is wider than the extremely narrow band associated with undamped resonance.

In general, any bends in the pipes should be as close as practicable to the inlet valve ports, blended smoothly into the straight sections, and their radii should not be less that 4 times that of the bore of the pipe. This arrangement leads to a minimum of both viscous losses and interference with the tendency for the air in the pipe to resonate freely.

11.5 The Helmholz resonator

Another device that has been applied to induction systems is the Helmholz resonator, Fig. 11.7. Because it is effective over only a very narrow band of frequencies, its use has been confined in the past to generating what has now become known as anti-sound, to eliminate induction pipe roar and exhaust boom. Anti-sound is of course a sound of the same frequency but opposite in phase to that which has to be eliminated. However, the principle of its application to turbocharged engines has been described in two papers presented before the IMechE in May 1990, at the Fourth International Conference on Turbocharging and Turbochargers. One is Paper C405/013, by G. Cser, of Autokut, Budapest, and the other is Paper C405/034, by K. Bsanisoleiman, of Lloyds Register, and B.A. French, of the Ford Motor Company. An earlier and equally interesting paper on this subject by Cser was C64/78, presented at the 1978 conference.

In general, Helmholz resonators have been used also to detect extremely faint noise signals. Another application is the damping of resonant vibrations, the damping effect being increased by placing porous material or cloth in the neck of the resonator. Also, it can be used to increase the sound pressure in an acoustic field at a particular frequency. This is of particular interest,

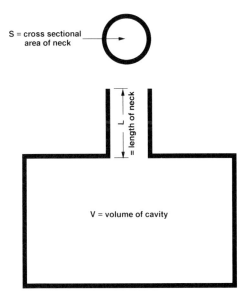

Fig. 11.7 Diagrammatic representation of a Helmholz resonator

because of its potential for enhancing the effectiveness of a tuned manifold. It can even act alone as a tuning device, though it has the disadvantage of being effective at only one frequency.

Application to induction pipes calls for one of two locations for the open end of the neck of the resonator. If is positioned at a displacement anti-node in the induction tract, it is in phase and therefore increases the amplitude of displacement of air in the tract. On the other hand, if placed at a displacement node, it tends to counteract the resonant vibration in the tract, because it is $\pi/2$ out of phase.

The Helmholz resonator generally comprises a short tube connected to an otherwise totally enclosed cavity. This cavity can be of any shape, though a bulbous form is preferred because it is less likely than almost any other to have natural modes of vibration that could influence the system as a whole. The air in the neck is assumed to act like a piston, alternately compressing and expanding that in the cavity. In other words, the air in the neck constitutes the mass, while the compressibility of that in the cavity forms the spring of a spring-mass system.

The wavelength of the vibrations it generates is large relative to the dimensions of the cavity. Its natural frequency f corresponds to the value of the angular frequency at which the reactance term disappears, and is therefore given by:

$$2\pi f = c\sqrt{(S/LV)}$$

i.e.
$$f = (c/2\pi)\sqrt{(S/LV)}$$

where c is the speed of sound, L the length of the neck, S the area of the neck and V the volume of the cavity. From the last term in the equation, it can be seen that the natural frequency increases as the square root of the area of the neck (mass of piston), and decreases as both the square root of the length of the neck (potentially available displacement of piston) and of the volume of the cavity (rate of the spring) are increased. Incidentally, provided the length of the tube is small relative to the wavelength of the sound at the resonant frequency, the effective length of the neck numerically is approximately the actual length plus 0.8 times S.

In acoustical applications, the Helmholz resonator is most effective at the lower end of the audible frequency range, of between about 20 Hz and 20 000 Hz. Expressed in terms of engine vibration, this is from about 1200 rev/min and, as regards incidence of inlet valve closure, from about 600 rev/min upwards, over the whole of the engine speed range.

ENGINE INDUCTION SYSTEM PHENOMENA

11.6 Interference between pipes, or charge robbery

The interference effect arises because the inertia of and depression in the air flowing into two adjacent cylinders during overlapping induction phases can cause one to draw in, from the manifold or plenum chamber, air that should be be going into the other. In fact, the filling of both cylinders is adversely affected.

Charge robbing potential increases with the number of cylinders, but decreases with engine speed. It is the reason for the troughs in the torque curves of many engines and, for example, illustrated by the lower curve in Fig. 11.25. With increasing speed, less time becomes available for the flow to reverse from one pipe into the other. Consider a typical engine in which the inlet valves are open from 12 deg btdc to 52 deg atdc, making a total of 242 deg. With the appropriate length of inlet pipe, this could produce something like the following lower limits below which inter-cylinder robbing could occur (Table 11.1):

Table 11.1

No. of cyls.	Deg overlap	Lower limit, rev/min
3	4	150
4	64	3900
6	124	4500

For induction system design purposes, the maximum, or limiting, velocity through the throat of the inlet valve is generally taken to be 100.6 m/sec (330 ft/sec). The mean velocity through the induction pipe is lower, mostly between about 1/3 and 2/3 of this value, though instantaneous velocities at points along the pipe may approach that of sound.

11.7 The inertia wave

The sequence of the pressure changes initiating the wave due to the closure of the inlet valve is as follows. First we have a condition of depression in the column of air as it flows into the cylinder. When its path is suddenly blocked by closure of the valve, it continues to flow towards the cylinder until it is suddenly brought to a halt by the now closed valve. We therefore have a local reversal of pressure from negative to positive. Bouncing back as an acoustic wave from the closed valve, a negative–positive pressure wave sequence travels towards the open end of the pipe, again at the velocity of sound, to be reflected back as a positive–negative sequence towards the cylinder. Therefore, the length of the induction pipe should be such that the positive pressure wave arrives back at the valve while it is open, so that the rate of flow into the cylinder will be significantly boosted.

The amplitude of the pulse increases directly with engine speed and, unlike stationary waves, it is not necessarily sinusoidal in form. Furthermore, the larger the volume, or mass, of air in the pipe, the greater will be the inertia effect and therefore the depression generated in the pipe by the downward motion of the piston. Consequently, the ratio of the induction pipe volume to piston-swept volume is an important criterion. Piston-swept volume is of course determined by other considerations.

The desirable length of pipe is determined by the engine speed at which optimum filling of the cylinder is required, so one is left with only the area as the independent variable. Two considerations limit the maximum diameter: one is the amount of space available and the other the practicability of tapering the tract down to the diameter of the inlet valve throat without incurring either manufacturing cost penalties or breakdown of the predominantly laminar into turbulent flow. Rectangular sections enclose larger volumes of air, but similar limitations apply to them, so circular sections are mostly preferred. An alternative is to taper the pipes, as already mentioned in connection with elimination of pipe end-effects. Tapered pipes do not necessarily have to be installed with their axes parallel, so they can be arranged relatively compactly alongside the head. Moreover, because they are tapered, it may be unnecessary for them to have trumpet ends.

11.8 Tuning the pipe to optimise the inertia wave effect

The inertia wave is longer and, as previously stated, its shape differs from those of the standing waves, Figs. 11.8 and 11.9. The time taken for it to travel from the inlet valve and back again is of course twice the length of the pipe divided by the velocity of sound ($2L/c$). In Fig. 11.8(a) is illustrated a characteristic curve of depression in the valve throat as a result of the motion of the piston during one complete revolution. The slight rise around tdc is caused by exhaust back-pressure during the overlap period. That around bdc is attributable to the inertia effect, resulting in a progressive build-up of pressure as the valve closes on to its seat.

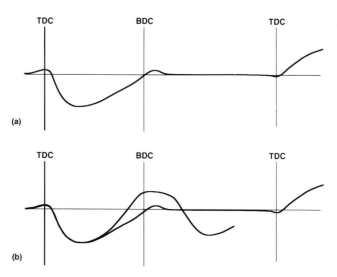

Fig. 11.8 At (a) is shown a characteristic curve of depression in the throat of an inlet valve while, at (b) a typical inertia wave is superimposed upon it

If the length of the induction pipe is such that the pressure wave arrives back at the inlet valve slightly earlier than the mid point of its open period, as in Fig. 11.8(b), the port pressure characteristics are optimum: if the pressure pulse were to arrive earlier, there could be a fall in pressure sufficient to cause reverse flow before the valve closes. The advantage of the later timing is that it increases the pressure difference across the valve port towards the end of the induction period, when it otherwise would be falling. It therefore improves the filling of the cylinder and tends to prevent reversal of flow as the inlet valve is closing. On the other hand, if the pulse pressure generated by this wave is high during the valve overlap period, scavenging of the clearance above the crown of the piston will be assisted.

Since this can happen at only one speed for any given length of pipe, and we are concerned with a reflection time requirement that varies with engine speed, it is more convenient to express time in terms of degrees of crankshaft rotation than in millisec. Another point is that we are concerned with timing relative to the period expressed in degrees of crankshaft rotation during which the valve is open. If we use the symbol t to represent time in millisec, it is convenient to use ϑ to represent time in terms of degrees rotation.

11.9 Tuning the pipe to optimise standing wave effects

Any pulse generated at the inlet end of the pipe will inevitably set up standing waves in the pipe. The time δ_t required for a single standing wave to be reflected back to its point of origin in terms of degrees rotation is twice the length of the pipe divided by the velocity of sound ($2L/c$). From the lower

OPTIMISING AIR INDUCTION—INDUCTION PIPE TUNING

diagram in Fig. 11.2, it can be seen that the wavelength of the fundamental frequency is $4L$, so δ_t is in fact half the periodic time.

During the time δ_t, the crankshaft rotates through an angle $\vartheta_t = 360N\delta_t/60$. If we substitute for δ_t, this becomes $\vartheta_t = NL/c$, where the suffix t refers to the time of the reflection, to distinguish it from ϑ_d, which is the time the valve is open, again expressed in degrees. It follows that if it were practicable for the single wave to be an exact fit in the induction period, it would occur when $\vartheta_t = \vartheta_d = 720/2n$, where n is the number of the harmonic or overtone.

If, in our calculations, we substitute the actual velocity of sound in the pipe for that of sound in free air, we have what might be termed an *induction wave front velocity*. Then perhaps the simplest way to exemplify the time for the wave front to travel one pipe-length is to assume a wave front velocity of 330 m/sec and a pipe length of 330 mm which, of course, will give a time of 1 millisec.

11.10 Harmonics of the standing waves

In addition to the standing wave at the fundamental frequency, harmonics are generated too by the initial impulse, Fig. 11.9, so a number of modes of vibration, superimposed on each other, occur simultaneously. Consequently,

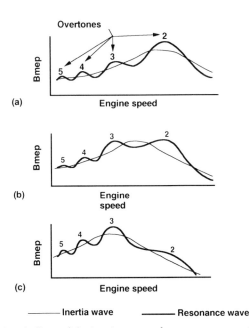

Fig. 11.9 The combined effects of the inertia wave and resonance, or standing, waves (a) with induction pipes tuned for maximum power, (b) to obtain a flat torque characteristic and (c) for torque back-up

the initial reflection at the fundamental frequency is accompanied by a ripple of reflections at the smaller wavelengths of the overtone frequencies The successive reflected pulses are of progressively smaller amplitudes owing to attenuation by viscous friction and out-of-phase reflections from bends and other obstructions in their paths. Consequently, those above the fifth harmonic are barely significant, and because they are closer together, the pressure peaks that arise from them are difficult to detect. In practice, the third and, possibly, fourth overtones, depending whether the valve timing is late or early, tend to be the significant ones. Long pipes and high speeds increase both the flow losses and the degree of attenuation of pulses.

The actual timing, relative to the depression wave, of the appearance of the succession of waves at the inlet port can be adjusted by advancing or retarding the opening of the valve. Neither the timing of valve opening nor the duration of overlap, however, have any significant effect on inertial ram, as distinct from resonance (or standing wave) ram, but they do affect exhaust assisted scavenge. Clearly, advantage could be gained by combining induction system tuning with variable valve timing, Chapter 10.

To fit the waves due to resonance into the valve open period, the following condition must be met: $n = \vartheta_t = \vartheta_d = 720/2n$, where n is the periodic time of the fundamental standing wave. If ϑ_t is less than $720/2n$, ripples will be superimposed on the depression pulse; if more, they may or may not affect the depression pulse.

Clearly, the inertia effect will be predominant at high speeds. This is because not only do the magnitudes of the pulses increase with speed but also, as the speed falls, the time available to fit more harmonics into the valve open period increases and, as previously mentioned, each successive wave reflection is weaker than its predecessor. Maximum amplitude of the standing wave occurs when the pipe length is such as to contain a single wave, which occurs when $L = \vartheta_t = \vartheta_d = 120$ degrees, and maximum overall amplitude is obtained when both the inertia and standing wave effects coincide.

Only the basic information has been given in this Chapter. In practice, the situation is further complicated by end-effects, those due to the presence of throttle valves and bends, the progressive motion of the closure of valves and by other factors. For more comprehensive and detailed information, the reader is advised to refer to articles by D. Broome, of Ricardo Consulting Engineers Ltd and papers by K.G. Hall of Bruntel Ltd. The former is a series in *Automobile Engineer*, Vol. 59, pp 130,180 and 262, while the latter were papers presented to the *IMechE and AutoTech 89*, Ref. C399/20. The latter contains a design chart presenting the graphical parameters in a manner such as to facilitate conceptualisation of an optimum induction system geometry.

In this chart, Fig. 11.10, the inertia parameter ϕ is plotted against the wave parameter θ, and a range of induction pipe diameters is superimposed. Wave parameter θ is proportional to N^2/L, where N = engine rev/min, and L the pipe length. The wave peak harmonics are shown as full and those of the wave troughs as dotted vertical lines. On the right, the scale of valve closing points T is equivalent to a range of ϕ from 3000 to 6000, but actual

OPTIMISING AIR INDUCTION — INDUCTION PIPE TUNING

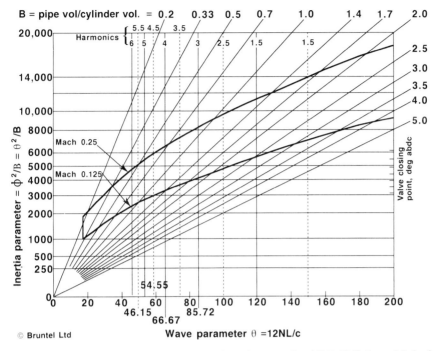

Fig. 11.10 An induction system tuning design chart, the copyright of K.G. Hall, Bruntel Ltd, of Maidenhead

values are not shown as they are empirical. They can be derived from data presented by Broome in *Automobile Engineer*, as quoted above. The Mach numbers of 0.125 and 0.25 approximate respectively to maximum torque and power, to take into account some of the scale effects. Ratios B, of volumes of pipes to cylinders, fan out in straight lines from the origin. As before c is the velocity of sound in the pipe.

PRACTICAL APPLICATIONS

11.11 Telescopic induction systems

Among the recent developments are the use of either telescopic or two-stage manifold pipes, with either a two-position control, to obtain optimum torque at two different points in the speed range or, in the case of the former, with an electronic management system varying it continuously. Another is the use of cluster manifolds.

Mazda, on their le Mans winning Wankel engine powered car, used electronically controlled two-position telescopic inlet pipes. A subsequent development by the same company was the use of the engine management

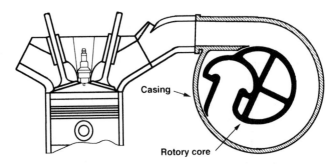

Fig. 11.11 The central portion, shown black, of the Tickford manifold is rotated to vary the effected length of the induction pipes. It extends the full length of the cylinder block to serve also as a plenum chamber

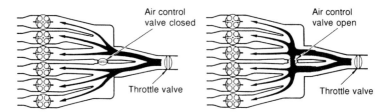

Fig. 11.12 The Toyota 7M-GE engine has a two-stage tuning layout which, in principle, is similar to that of the GM Vauxhall-Opel Dual Ram system described in Vol.1

system to vary the pipe length continuously. However, telescopic pipes, though ideal in some ways, have severe disadvantages. Their control mechanisms are complex and can be difficult to accommodate, and reliability over long periods in service would be difficult to ensure. The Tickford system however, Fig. 11.11, should be both more reliable and easier to install. This is similar in principle, but has an approximately toroidal telescopic duct. The central portion, which also forms part of a cylindrical plenum chamber common to all the cylinders, is rotated to vary the length of the individual telescopic sections serving the cylinders.

11.12 Vauxhall and Toyota two-stage induction pipes

Outstandingly good results have been obtained with the Vauxhall/Opel Dual Ram system described in Vol. 1, and that of the Toyota 3 litre DOHC, 7M-GE 6-cylinder engine, Fig. 11.12. Both are similar in principle: at high speeds, six primary pipes are in operation and, at lower speeds, their lengths are extended by switching them into tandem with two secondary pipes. The change-over point of the Toyota system is effected at 4200 rev/min, by opening or closing the valve in the dividing wall of the plenum chamber into which the air from the twin secondary pipes enters, and from which it continues into the primary pipes. That of the GM system is effected at 4000 rev/min

OPTIMISING AIR INDUCTION — INDUCTION PIPE TUNING

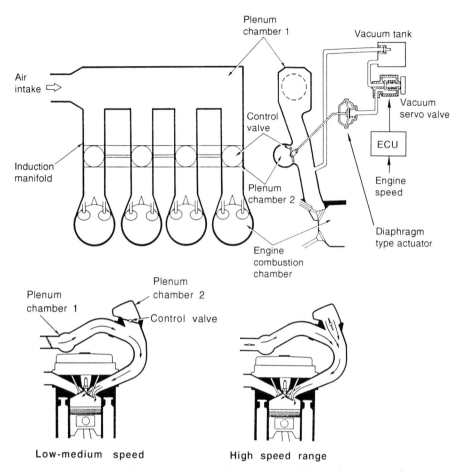

Fig. 11.13 Toyota ACIS system for the 3S-GE engine. The ECU controls a vacuum valve serving a diaphragm type actuator for the butterfly type control valves in a throat between the induction pipes and a secondary plenum chamber

by actuating a similar valve which, when closed, divides the intermediate plenum chamber in two, each half forming an interconnecting duct between three primary pipes and one of the two secondary pipes.

In both systems, the valve is open at high speeds, so the effective length of the induction system is that from inlet valves to plenum chamber. When this valve is closed, for operation at speeds below the change-over point, the effective length is that from the inlet valves to the throttle valve.

Subsequently, Toyota introduced, for their 4-cylinder 2 litre DOHC 3S GE engine, an entirely different arrangement, Fig. 11.13, which they term the Acoustic Control Induction System (ACIS). Again it comprises primary and secondary pipes, which together constitute one long pipe for each cylinder. At speeds above 5100 rev/min, they are in effect shortened by opening butterfly valves approximately mid-way between their outer ends and the inlet valves. These butterfly valves are in throats leading to a plenum chamber which is common to all four pipes.

11.13 Peugeot obviate inter-cylinder interference

Peugeot, on V6 engine for their 605 model have the Variable Acoustic Supply Characteristics (VASC) system, Fig. 11.14, comprising three stages. The illustration shows the system diagrammatically: in practice, to render it more compact, the portion on the right that houses the two control valves is turned up through 90 deg and the remainder then turned a further 90 deg, to overhang the two straight portions of the plenum chamber, Fig. 11.15. A third control valve is situated in a passage interconnecting these two straight portions, and between them is an accumulator, for storing the manifold depression needed for actuating the three control valves. Overall control is effected by the Peugeot Fenix 4 electronic control unit.

Air passing the two throttle valves on the left enters the two arms of the plenum chamber and, at all speeds, passes directly into the individual induction pipes serving the six cylinders. According to the manufacturers, at speeds below 4000 rev/min, a high velocity of flow is needed and this is obtained by reducing the effective length of the plenum chamber by closing all three control valves. In this condition each of the two main plenum chambers serves one cylinder block.

At speeds between 4000 and 5000 rev/min, say the manufacturers, the air needs to flow smoothly between the arms of the plenum chamber, so the

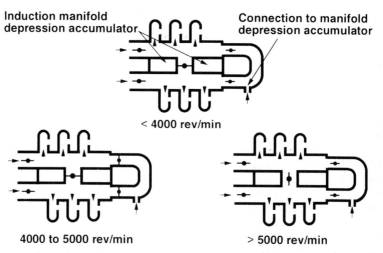

Fig. 11.14 Diagrammatic illustration of the principle of operation of the Peugeot VASC system. The port in the U-shape portion is a connection to the manifold depression reservoir, which serves the diaphragm actuators for the control valves

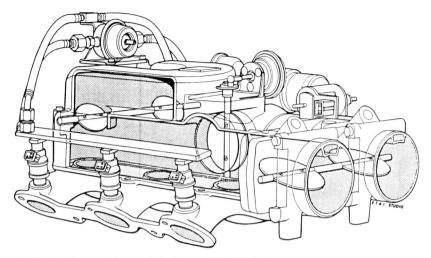

Fig. 11.15 The actual layout of the Peugeot VASC induction system

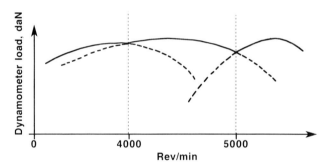

Fig. 11.16 Curves of test bench brake load plotted against speed for the three conditions of operation of the Peugeot VASC induction system

two valves at their ends remote from the throttle valve are opened. Then, above 5000 rev/min, to obtain optimum breathing, the third valve, directly interconnecting the arms, is also opened.

The explanation just given would appear to be an over-simplification. Presumably, the three-stage increase in the plenum chamber length is to avoid inter-cylinder robbery. This appears to be borne out by the torque curves illustrated in Fig. 11.16.

We might perhaps analyse this system by comparing it with Figs. 11.24 and 11.25. Then, at speeds below 4000 rev/min, the pipe length would be the equivalent of Ld_1, between 4000 and 5000 rev/min it would be Ld_{1-2}, while above 5000 rev/min it would be Lp_2. If the distance needed between pipe ends to avoid interference between cylinders is L_I, the engine must operate above a speed N_I. In a V6 engine, the cylinders in which induction strokes are overlapping are generally in opposite banks so, even with conventional manifolding, the distances are generally relatively long.

If the period of the cam is 242 deg, the overlap ϑ between the induction strokes of any two cylinders during one stroke of a piston (180 deg) is 62 deg. We can use the formula for the time, in degrees of crankshaft rotation, required for an inertia wave to travel away from and be reflected back to an inlet valve, for calculating the time for a pulse to travel from the end of one port to another. This formula is, $\vartheta_I = NL/c$ but, in calculating potential for interference, the length L_I is half that of L for a reflected wave in a pipe and, since the speed of sound is in m/sec, the units of N must be in deg/sec. Consequently, the calculation is as follows:

The interference angle (or overlap) $= \vartheta_I = \dfrac{360\,N_I}{60} \times \dfrac{2L_I}{c}$

or

$$\vartheta_I = \dfrac{12 N_I L_I}{c}$$

Where L_I = half the distance between the inlet valves of the interfering cylinders.

If we take the velocity of sound to be 330 m/sec and $\vartheta = 62$ deg, we have:

$$N_I/L_I = \dfrac{330}{12} \times 62 = 1075$$

Over the three speed ranges quoted for the Peugeot engine, the lengths L_I become:

Engine rev/min	<4000	4000–5000	>5000
L_I mm	550	450	350
N_I, rev/min	3100	3790	4780

Note: Actual and calculated values will differ if c in the induction pipes is not 330 m/sec, as explained in Section 11.4.

11.14 The Honda variable volume induction system

Honda's Variable Volume Induction System (VVIS) on their NSX 3 litre V6 engine is a two-stage arrangement. The illustration, Fig. 11.17, shows a secondary chamber beneath the main plenum chamber, mid-way between the cylinder banks. The three inlet pipes from each cylinder bank rise from the inlet ports, and pass immediately over the secondary plenum chamber to enter the sides of the main chamber. Near the upper end of each pipe and interconnecting it with the secondary chamber is a port, opened and closed by manifold-depression-actuated valve.

At speeds above 4800 rev/min, the effective length of the induction pipes is that between the inlet valves and the main plenum chamber. Below that speed, the manifold depression actuated valves open all six pipes simultaneously to the secondary plenum chambers, while also shortening

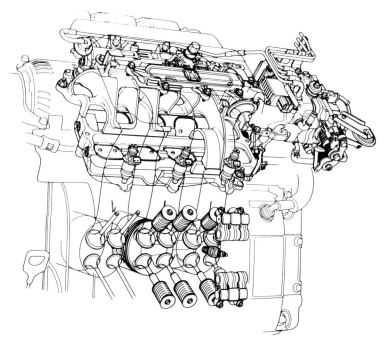

Fig. 11.17 The Honda Variable Volume Induction System for the V6 NSX engine has a butterfly control valve interconnecting each induction pipe with a secondary plenum chamber, as also does the Toyota ACIS. On the lower bank of cylinders can be seen the variable valve timing electric control mechanism (VTEC) described in Chapter 10

their effective lengths. This would appear to increase their resonant frequency while, at the same time, increasing the active volume of plenum chamber to reduce any tendency to inter-cylinder robbery of charge. Performance curves for the open and closed valve conditions are illustrated in Fig. 11.18. This shows that what would have been a flat portion of the curve in the run up to the peak has been lifted to give an extended range of high torque from about 3500 to 6500 rev/min. In this instance, however, the improvement has been achieved by a combination of VVIS with what Honda term Variable valve Timing and lift Electronic Control (VTEC).

11.15 Volvo siamesed pipe system

On the Volvo 2.0 litre 5-cylinder in-line engine for the 850 GLT, the throttle barrel is bolted on to one end of a cylindrical plenum chamber alongside the cylinder block. Five individual primary pipes issue from the base of the plenum chamber, turning through about 315 deg up and over the plenum to join the inlet ports, Fig. 11.19. Each pipe comprises a pair of siamesed airways, its cross section resembling a figure eight. The top loop of the figure

296   OPTIMISING AIR INDUCTION — INDUCTION PIPE TUNING

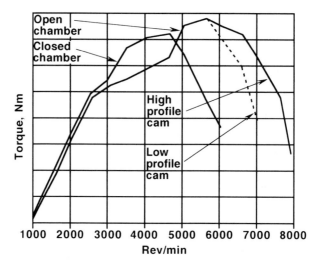

Fig. 11.18 The two curves on the left show the effect on torque of opening and closing the control butterfly valves of the Honda ACIS system, while the extension on the right shows how the speed range is increased by application of Honda's variable valve timing mechanism

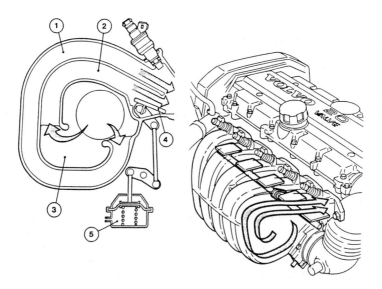

Fig. 11.19 Right, the Volvo V-VIS variable induction system with, left, a diagrammatic view of a vertical section through the first induction pipe
1 Long pipe, 2 short pipe, 3 plenum chamber, 4 control valve, 5 diaphragm type actuator

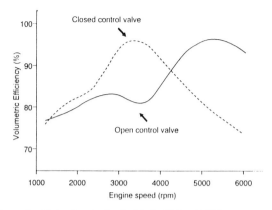

Fig. 11.20 Volumetric efficiencies obtained with the Volvo V-VIS system

eight is of slightly smaller diameter and its airway about twice as long as the lower loop. A flap valve over the upper end of the lower loop is either fully opened or closed by a diaphragm type actuator subjected to manifold depression and controlled by the ECU.

At speeds below 1800 rev/min, both ducts are open, providing air flow capacity for acceleration. Between 1800 and 4200 rev/min, but only so long as the throttle is 80 deg or more open, the shorter duct is closed. Above 4200 rev/min, both ducts are open again to afford maximum flow potential. The efficacy of the arrangement is apparent from Fig. 11.20.

11.16 Some interesting alternatives

On the basis that the most effective reflection is that of the inertia wave, some interesting alternative systems have been proposed, and in some instances put into practice, by K. G. Hall, of Bruntel Ltd. For example, pointing out the potential for inter-cylinder robbery with conventional manifolds having short branch pipes, he favours replacing such systems with a cluster of pipes in a conduit which, by sealing its ends around the pipes, is made to serve also as a plenum chamber. Alternatively, a plenum chamber can be simply a box surrounding the four pipes about mid-way between their ends.

In either case, there are ports in one side of each pipe in the plenum chamber, which can be opened and closed by, for example, a single sliding valve. When this valve is open, the pipes are in effect shortened for operation at high speeds and, when closed, the whole length (primary plus secondary) of the pipes is effective for optimising bmep at lower speeds.

When the sliding valve is open, the air can flow in through all four pipes, and three of them are free to feed through the plenum chamber into the fourth and thence into the cylinder in the induction phase. This has the advantage that velocities of flow in the outer half of the induction system

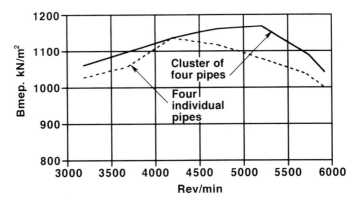

Fig. 11.21 Computer simulation comparing the bmep and torque characteristics of a conventional individual induction pipe system with those of a Bruntel cluster arrangement, the latter being significantly better in the higher speed range

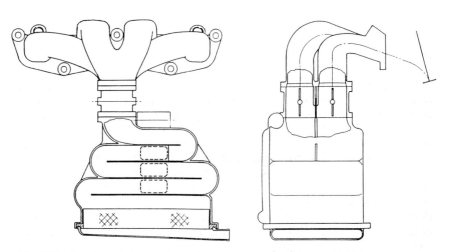

Fig. 11.22 The Bruntel zig-zag cluster pipe arrangement for three-stage tuning

are low and therefore so are the losses at high engine speeds. At lower engine speeds, velocities are of course in any case lower. Because the valve is in the plenum chamber, sealing the ports when it is closed is not critical. The computer simulated curves in Fig. 11.21 compare the bmep obtainable with this two stage tuning of a cluster with that obtained with a group of four separate pipes tuned for maximum bmep at 4000 rev/min.

A further development of the cluster pipe theme for a four cylinder engine, but providing three-stage tuning, is illustrated in Fig. 11.22. The inner pair of branch pipes of the manifold are led into one throttle barrel and the outer pair into another. From this point, two larger diameter pipes zig-zag down

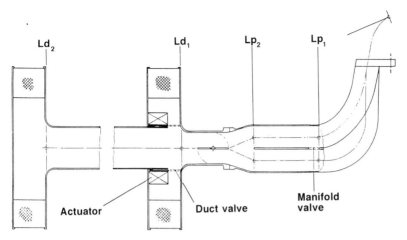

Fig. 11.23 A ducted uniflow induction system controlled by means of two valves, but providing a bmep characteristic similar to that of the more complex system of Fig. 11.22

to a plenum chamber, which totally encloses them and houses also the air filter.

In this instance, the primary section comprises the manifold and twin throttle barrels, and the secondary section is the zig-zag portion. With the conventional firing order, there is no inter-cylinder robbery. A single actuator, in the form of a sliding plate with ports appropriately arranged in line can be designed to open and close the ports in the zig-zag secondary portion.

A simpler method of providing multi-stage tuning, giving a similar torque characteristic, is illustrated in Fig. 11.23. The induction system is divided into a pipe cluster, which is the primary manifold, and a duct incorporating two filters, which is the secondary duct. In the primary manifold is a valve that can be opened to shorten its effective length from Lp_2 to Lp_1. The length of the secondary duct can be reduced from Ld_2 to Ld_1 by lifting the outer section to open up a gap between it and the next one.

For high speed operation, the manifold valve is open so that the effective length is Lp_1. Even when this valve is open, however, the system will automatically tune to the length Ld_1 as the speed falls. This effect is similar to that when a player blows with increasing or decreasing force into certain of the wind instruments: the note suddenly switches respectively to a higher or lower harmonic.

When both the manifold and duct valves are closed, the system tunes to Ld_2 in the low speed and Lp_2 in the higher speed modes of operation. From Fig. 11.24, in which the resonating lengths are indicated above the characteristic torque curve, it can be seen that by virtue of valve operation and automatic tuning, the torque can be optimised over the whole of the speed range. An indication of how the various resonances lift the torque curve is given in Fig. 11.25. Incidentally, this system can be relatively easily

300 OPTIMISING AIR INDUCTION—INDUCTION PIPE TUNING

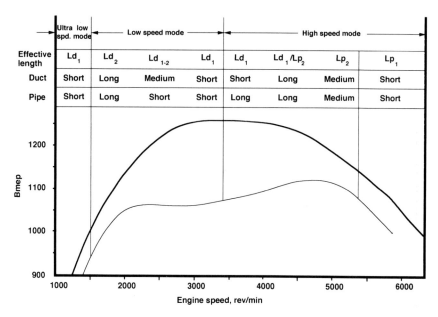

Fig. 11.24 Bmep curve obtained with the system in Fig. 11.23 with, listed above it, the effective primary pipe and secondary duct lengths

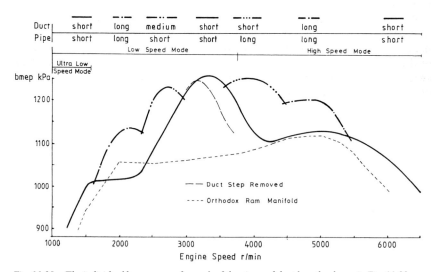

Fig. 11.25 The individual bmep curves for each of the pipe and duct lengths shown in Fig. 11.23

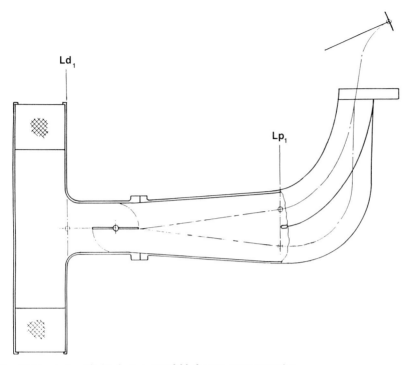

Fig. 11.26 A simplified induction manifold, for two-stage operation

adapted for the use of only one filter, without altering the sequence of resonances.

A simpler adaptation of this system, but for only two-stage operation, is illustrated in Fig. 11.26. At high speeds there is a strong tendency for the pressure waves to be reflected back from the ends of the short curved sections of the pipe to the inlet valve. As the speed falls, the reflections automatically switch to the outer end of the secondary pipe, so that the longer wave mode of vibration can come into resonance.

Chapter 12

Optimising air induction — turbocharging, and the Comprex pressure wave charger

Legislative pressure for reduction of emissions is the major factor leading to the current rapidly increasing application of turbocharging to diesel engines. There is also a small benefit, perhaps about 3%, in terms of reduced fuel consumption. With rising load, and therefore fuelling rate, on a naturally aspirated engine, the temperature and pressure of the charge increase. Consequently, so also does the energy completely lost to the exhaust. By incorporating a turbo-compressor combination to convert a significant proportion of this energy for supplying correspondingly increasing quantities of air, noise and exhaust gas emissions, including particulates, can be reduced and torque enhanced.

Further benefits are obtainable by aftercooling. These include reduction of NO_x, increased thermal efficiency, lower engine temperatures, and improved reliability. In short, turbocharging with aftercooling helps the diesel engine to compete with its gasoline fuelled counterpart, as regards specific power output in terms of both kW/litre and, to a lesser degree, kW/kg. The progress that can be made by the application and development of turbocharging and aftercooling is exemplified by the work done by Scania, who have produced the graph illustrated in Fig. 12.1. This illustration also shows compares these results with those obtained with indirect injection and turbo-compounding; the latter is briefly mentioned at the end of Section 12.14.

With natural aspiration the air supply initially increases approximately linearly with rotational speed of the engine, and then the rate of increase falls off as speed increases. Turbocharging, on the other hand, can be utilised not only to match the air supply more closely to the torque, and therefore fuelling rate, demanded but also to shape the torque curve. Many more advantages, to be elaborated upon in this chapter, also arise, and all add up to a good reason for the widespread adoption of turbocharging rather than supercharging.

Superchargers suffer the disadvantages of requiring a mechanical drive and, in most instances, also of being bulky and noisy. On the other hand, the output of the positive displacement types is approximately proportional to the rotational speed of the engine and therefore to its inherent air

OPTIMISING AIR INDUCTION — TURBOCHARGING

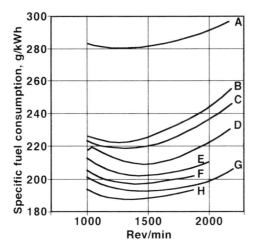

Fig. 12.1 These curves show the progress that has been made in reducing specific fuel consumption of the Scania diesel engines over approximately half a century since the early 1940s. This is fairly typical of what has been happening throughout Europe. No turbo-compound engine has gone into production for an automotive application, but some are to be found in industrial applications

Year	Engine	Type	Output, kW/dm^3
1944	D 600	Pre-combustion chamber	11
1949	D 622	Direct injection	12
1960	DS10	Turbocharged	15
1974	DS11	Turbocharged	19
1980	DS14	Turbocharged	20
1982	DSC14	Aftercooled	22
1984	DSC9	Aftercooled	24
1991	DTC11	Turbocompounded	27

requirement. Consequently, they tend to be suitable mainly for two types of application: the first is for gasoline engines, since the fuel and air are introduced ready mixed and control is effected by throttling, while the second is for improving the scavenging of two-stroke engines. Even so, turbocharging is generally preferred for gasoline engines, mainly because it does not require a mechanical drive. Since superchargers are rarely used on automotive diesel engines, the space saved by ignoring them in this chapter will be devoted to explaining some aspects of design and application relating to the performance of turbocharger and its matching to the engine.

12.1 Turbocharging

Although turbochargers have been available before World War II, it was not until about 1957 that they had been developed to an acceptable degree of efficiency and reliability, and at a low enough cost to be seriously considered for application to heavy commercial vehicles. Subsequently, smaller versions suitable for lighter vehicles, including cars, have been introduced progressively. For the large medium and low speed industrial diesel engines, both radial inflow and axial flow turbine and compressor combinations are used, but the axial flow types are not favoured for automotive applications.

Fundamentally, turbocharging enhances the efficiency of the diesel engine in four ways. First, it increases the mean effective pressure and density of the charge and therefore the overall efficiency of the power unit. Secondly, the combination of high air:fuel ratios and mean effective pressures produces good indicated efficiencies. Thirdly, the increased air:fuel ratios tend to lead to lower exhaust gas temperatures. Fourthly, as previously indicated, the turbocharger is driven by exhaust gas energy that otherwise would be wasted, and suffers neither the installation problems nor the parasitic losses associated with a mechanical drive. For all these reasons, it has been widely adopted for both commercial vehicles and cars, especially those that are diesel powered.

On the other hand, high degrees of turbocharging tend to imply high inlet air temperatures due to both the compression of the air and heat transfer from turbine to compressor. The outcome is high combustion temperatures, which tend to cause both NO_x formation and increased heat losses to coolant. Aftercooling of the compressor output before it is delivered to the cylinder is therefore highly desirable, and especially with compressor pressure ratios of over about 1.5:1. Incidentally, aftercooling is defined as cooling the charge after it has passed through the compressor, while intercooling is cooling it between two compressor stages, the latter being virtually unknown in road vehicle applications.

In the past, air-to-liquid was preferred to air-to-air aftercooling, mainly because it was more compact, even though it might entail placing a large heat exchanger in front of that used for cooling the engine. However, it suffers two disadvantages: first, the temperature differential between the liquid coolant and air tends to fall as the load on the engine increases, reducing its effectiveness; secondly, the plumbing needed adds to the weight and complexity of the installation, and is a potential source of failure. More recently, however, some extremely compact and highly efficient air-to-air aftercoolers have been introduced, notably that installed on the Ford Mondeo, Fig. 12. 2, launched early in 1993. It seems likely, therefore, that this type of system will now gain ground.

In general, engine efficiency increases by about 0.5% for every 10 deg C fall in charge temperature. Given that the air:fuel ratio remains constant, the outcome is a corresponding increase in power output by about 3.5%. Furthermore, by reducing the temperatures of pistons and other components, charge cooling allows the power rating of the engine to be increased.

Despite its many advantages, however, turbocharging also has some disadvantages. Relative to natural aspiration, it has a relatively narrow useful boosted speed range and, relative to positive displacement supercharging, it offers poor low speed torque. Moreover, because turbochargers run up to maximum speeds ranging from 100 000 to well over 150 000 rev/min, they take a significant time (in extreme instances as long as 2 to 3.5 seconds) to overcome the rotational inertia. Consequently, if rapid acceleration from low

OPTIMISING AIR INDUCTION — TURBOCHARGING

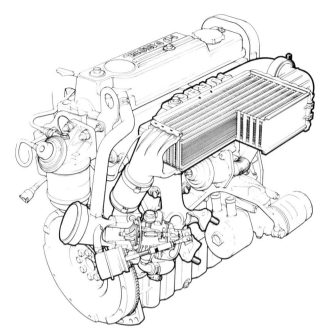

Fig. 12.2 The Ford Mondeo 1.8 litre turbo diesel engine has an extremely compact air-to-air charge cooler

to maximum speed is demanded, the engine tends to be slow to respond. A great deal of research and development therefore has been, and continues to be, aimed at reducing the inertia of the rotating assembly to a level acceptable to drivers. This aspect is covered in more detail in Section 12.12.

The detail design of turbochargers is a highly complex and specialised science, for more information on which readers are advised to refer to one of the text books. Among these is 'Turbocharging the Internal Combustion Engine', by N. Watson and M.S. Janota, Macmillan, 1982. Another is 'Supercharging of Internal Combustion Engines' by K. Zimmer, Springer Verlag, which was published in the English as well as German languages in 1978. The latter covered the fundamental principles of both supercharging and turbocharging equally thoroughly. More recently, an extremely good summary of the practical design considerations has been given in Paper C405/024, presented by Flaxington and Mahbod, of Allied-Signal Garrett Automotive Group, in the Proceedings of the IMechE International Conference on Turbocharging and Turbochargers, May 1990, ISBN 0-852987196.

12.2 The turbocharger unit

As previously stated, axial flow turbines and compressors are suitable for only very large engines. Another feature of the large industrial engines, and

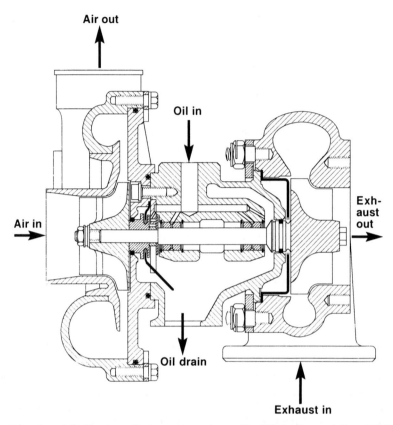

Fig. 12.3 One of the K series turbochargers manufactured by Kühnle, Kopp und Kausch (KKK). This version has a single entry turbine with a tapered nozzle, and its compressor diffuser has parallel sides without vanes. Note the sturdy thick-wall bearings designed for very high speed running, and the large drain in the base to ensure that the lubricant can leave the very hot casing with as little delay as possible

one that has been only rarely applied to automotive applications, is the delivery of the exhaust gas into a plenum chamber before passing it on to the turbine and only then converting its pressure into kinetic energy. This is termed *constant pressure* turbocharging. For automotive engines, *pulse* turbocharging, in which exhaust gas pulse energy is utilised, is the system mostly adopted.

Inside the casing of a conventional automotive turbocharger, Fig. 12.3, are three rotating parts. One is the shaft, which is carried in a bearing assembly approximately mid-way between its ends. Mounted on one end of this shaft is the rotor of the *radial outflow compressor* and, on the other, that of the *radial inflow turbine*.

Exhaust gas, containing both pressure and kinetic energy, is delivered into a volute, from which it passes radially inwards through a peripheral nozzle to drive the turbine, to power the compressor. The cross section of the volute passage decreases progressively to help not only to distribute the gas inflow uniformly round the periphery of the rotor but also to convert the pressure into kinetic energy. A proportion of this energy is imparted to the turbine rotor and transmitted by the shaft to the compressor. On reaching the hub the gas, now at reduced energy, is deflected through 90 deg, to flow axially out into the exhaust down pipe.

Air is ducted through a central axial intake into the compressor the vanes of which impart energy to it while forcing it radially outwards. It then passes through a peripheral diffuser and volute, into the induction manifold.

Sturdy precision plain bearings designed and developed specifically for the very high speeds of rotation experienced are mostly used for carrying the shaft, though development of suitable rolling element bearings continues because of the prospect of reduced friction. On the other hand, rolling element bearings are of course heavier and less compact. Lubrication is important as a means of cooling plain bearings but, in rolling element bearings, it has to be strictly controlled to avoid excessive oil drag at high speeds. Precision and close tolerances are important, especially in plain bearings, for avoiding, for example, the tendency, which can arise in certain circumstances, for their shafts to roll around inside the bearing instead of rotating coaxially within them.

12.3 Turbine nozzles and compressor diffusers

Turbine nozzles and compressor diffusers have much in common, though the former of course have to survive much higher temperatures. Either the nozzle or the diffuser, or both, can take any of several forms. The diffuser of the Holset compressor, Fig. 12.4, has a parallel nozzle. With such an arrangement it is relatively easy to insert vanes or other devices for converting the kinetic into pressure energy or vice versa. In other instances the sides are inclined to form tapered peripheral diffusers or nozzles, as in the twin entry turbine of the same turbocharger. Hitherto, vaneless nozzles and diffusers have been preferred because they are less costly, easier to produce, and generally provide a broader performance map Section 12.8.

The reason why vanes increase the efficiency of turbines, because they improve both the distribution of flow of the gas around the rotor, and its angle of incidence as it enters it, Fig. 12.5. However, this higher efficiency is obtained over a narrower range of mass flow. Diffuser vanes in compressors, on the other hand, reduce the effective speed range of the turbocharger, owing to an increased tendency to stall and therefore surge as the gas throughput falls below its design rate.

12.4 Potential for improving turbine efficiency

Although vaned are more efficient than vaneless turbine nozzles, each vane is an extra part that can suffer erosion and fatigue damage, and thus lead

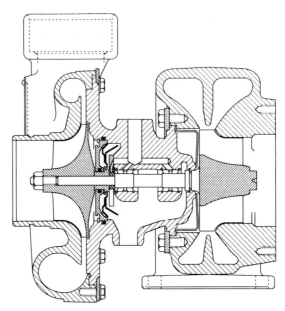

Fig. 12.4 This Holset turbocharger has a twin entry turbine with tapered nozzles while its compressor diffuser has parallel sides. Note the gas sealing arrangements at both ends, the heat shields and the oil deflector ensuring that lubricant leaving the bearings cannot be drawn into the compressor

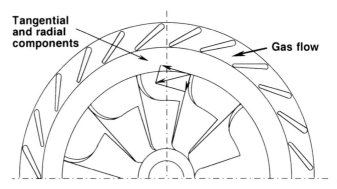

Fig. 12.5 Turbine nozzle vanes and associated air flow vectors

to unreliability and shortened life. Consequently, in the past there has been little incentive to accept the extra cost associated with vanes.

Even so, serious attention is being devoted to further development of vaned nozzles and especially variable geometry mechanisms for automatically varying vane angles, to broaden the range of mass flows over which the efficiency remains high. Incidentally, this trend is apparent in respect of both turbine nozzles and compressor diffusers. However, the benefit to be gained by introducing variable geometry vanes into the diffuser in a compressor is much greater, so it is in this area that the most progress has been made.

The ultimate aims of varying the geometry are reduction of both specific fuel consumption and exhaust gas emissions. Variable geometry vanes are of course easier to accommodate and install in parallel sided than tapered nozzles. An interesting example of a swinging vane type variable nozzle in a turbine is the Honda Wing turbo system, Figs. 12.6 and 12.7. A simpler alternative to varying the angles of the vanes is to narrow and widen respectively the throat of a vaneless nozzle, by moving one of its side walls inwards and outwards.

In general, pressure drop across a turbine (the expansion ratio) is a function of the square of the mass flow. Turbine power output, on the other hand, is a function of both the mass flow and the expansion ratio. Consequently, turbine power, and therefore the speed at which it drives the compressor, increases more rapidly than mass flow. This is significant for matching the turbine to the engine, Section 12.13.

The turbine pressure ratio is small relative to the expansion ratio in the engine cylinders. Consequently, the temperature of the gas leaving the turbine is still high. Therefore the energy in it could be exploited by adding a separate turbine stage for delivering torque directly to the engine output shaft, Fig. 8.7. However, although such devices are being investigated, they have so far proved to be too costly, bulky and heavy for automotive applications.

12.5 The compressor

During the passage of the air through the compressor, the energy derived from the turbine is imparted to it by the blades of the rotor. This accelerates the air to a high velocity, with both tangential and radial velocity components, before it is discharged from the periphery. Finally, to convert the velocity energy into pressure energy, it is expanded through a diffuser which, as previously indicated, may or may not be vaneless.

The speed of rotation of the turbine and compressor assembly is at all times determined by the balance between the torque output of the former and the torque absorbed by the latter. Therefore, the pressure ratio across the turbine is, similarly, automatically balanced against that across the compressor.

In most automotive applications, the torque transmitted to the compressor is relatively small, generally rising to a maximum of perhaps about 10–15 Nm. Heat transfer from turbine to compressor is restricted by the presence between them of the bearing assembly and heat shielding in various forms ranging from simply the gap between their casing walls to actual thermal shields. Water cooling is sometimes used, but it is complex, costly and introduces the possibility of leaks and other troubles.

While engine designers tend to prefer to think in terms of the boost pressure of an engine, the more widely used criterion of performance of a turbocompressor is the ratio of its input pressure to output pressure, the

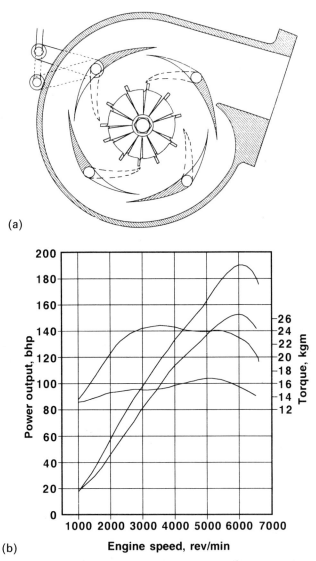

Fig. 12.6 (a) The Honda Wing turbine with the wings shown in the aligned (full lines) and fully deflected (dotted lines) positions; (b) performance curves of the Wing turbocharged and naturally aspirated versions of the Honda 2 litre V6 24 valve + PMG-FI engine

pressure ratio, of the compressor. In general, maximum pressure ratios actually utilised in high speed diesel engines range around 2:1, the trend being currently upwards. For some medium and low speed diesel engines, however, pressure ratios significantly higher than 3.5:1 are encountered. Detonation, mechanical and thermal loading, especially of exhaust valves,

OPTIMISING AIR INDUCTION—TURBOCHARGING

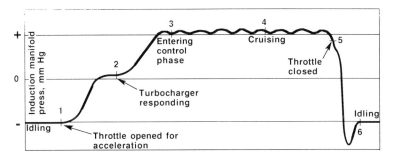

Fig. 12.7 This is a plot of the induction manifold pressure of the Honda engine running under full load. It starts at 1, where the turbo wings are in the position indicated by the full lines in Fig. 12.6, and ends at 6 where they are returned to the same position. Stages 2 to 5 represent the depression as the turbo wings are opened progressively in four equal increments between these two positions

Table 12.1 THE TURBOCHARGER DATA FOR THE ROVER SPARK IGNITION RANGE

Vehicle	Rover 800 Vitesse	Rover 200/400 Turbo
Engine size, cm^3	1994	1994
Max power, kW	134 kW	149
@ rev/min	6000	6000
Max torque, Nm	216	240
@ rev/min	2000–5500	2000–5500
Max. boost, kPa abs.	173	180
Turbo model	Garrett T25 BCI 15	Garrett T25 BCI 15
Max. press. ratio obtainable from unit	2.2:1 (220 kPa)	2.2:1 (220 kPa)
Ratio of max. boost: max. pressure ratio obtainable	1.87:1	1.96:1

and even the torque capacity of the transmission, tend to limit the degree of boosting acceptable for spark ignition engines, so the pressure ratios utilised tend to be lower.

Since the Rover Group produces a range of vehicles powered by both diesel and spark ignition turbocharged engines, Tables 12.1 and 12.2, we can take it as being fairly typical of current practice. The 1994 Saab range, Table 12.3, all except the Aero equipped with Mitsubishi T25 turbochargers, are of interest too because this Company offers some lightly turbocharged versions matched to provide enhanced torque, and therefore flexibility, at the lower end of the speed range. This is advantageous in cars operating in urban traffic, especially if their performance at the upper end of their speed range is in any case high without turbocharging.

The map illustrated in Fig. 12.9 is exceptionally narrow because of the high pressure ratio attainable with that particular turbocharger. Performance

Table 12.2 THE TURBOCHARGER DATA FOR THE LAND ROVER DIESEL RANGE

Vehicle	Discovery 200 Tdi	Defender 200 Tdi	Discovery and Defender 300 Tdi
Engine size, cm^3	2495	2495	2495
Max power, kW	83	80	83
@ rev/min	4000	3800	4000
Max torque, Nm	265	255	265
@ rev/min	1800	1800	1800
Max boost, kPa abs.	181	181	202
Turbo model	Garrett T25 comp. T2 turbine	Garrett T25 comp. T2 turbine	Garrett T25 T2 turbine
Max. Press. ratio obtainable from unit	2.5:1 (250 kPa)	2.5:1 (250 kPa)	2.5 (250 kPa)
Ratio of max. boost: max. pressure ratio obtainable	1.724:1	1.724:1	1.81:1

Table 12.3 PERFORMANCE DATA FOR SAAB TURBOCHARGED RANGE OF SPARK IGNITION ENGINES

Vehicle	900 coupé	9000 E'power	9000 E'power	9000 Turbo	9000 Aero†
Engine, cm^3	1985	1985	2290	2290	2290
Max. power, kW	115	115	130	153	171
at rev/min	5500	5500	5700	5500	5500
Max torque, Nm	263	210	260	323	342
Comprn. ratio	9.2:1	8.8:1	9.25:1	9.25:1	9.25:1
Boost, bar:					
Basic	0.4	0.4	0.4	0.40	0.45
@ rev/min	3000	3000	3000	3000	3000
Maximum	0.79	0.59	0.40	0.81/1.0*	1.08

† Automatic/manual transmission.
* With TDO 4HL15G turbocharger.

maps of most turbochargers designed for maximum attainable pressure ratios of up to no more than about 2.5:1 tend to be as much as twice as wide. Moreover, if the engine is turbocharged at relatively a low pressure ratio, it in any case operates mostly in the widest part of the performance map.

The boosting effect of the compressor, Fig. 12.8, varies with mass flow rate, which is a function of the square of its rotational speed. Maximum rates are of course determined by the sizes of the passages through the unit, and therefore their choking effect. At low rates of flow, yet another limit is imposed: this is a tendency for the flow to *surge*.

12.6 Compressor surge

Surge is a sudden reversal of flow from the high pressure delivery to the low pressure intake port, and it occurs cyclically. As the velocity, and therefore

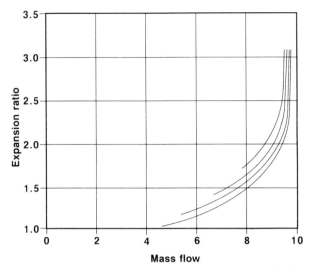

Fig. 12.8 Turbocharger characteristics for a range of speeds, increasing from bottom to top. The pressure drop across the turbine (expansion ratio) is a function of the mass flow through it, and its power is a function of both

kinetic energy, of the flow falls, a condition arises in which that energy is insufficient to overcome the adverse pressure gradient between the compressor inlet and outlet. At this point, the flow collapses, causing the pressure gradient to die away abruptly. Consequently, the flow is re-established, so the pressure gradient builds up again until the cycle begins again.

Operation in the surging condition can not only interrupt combustion in the engine cylinders but also even cause mechanical failures. Surge mostly occurs throughout the unit and should not be confused with *stall*, which is a totally different phenomenon, Section 12.7, but nevertheless one which can trigger surge. Both can be noisy, though surge is usually the louder of the two.

12.7 Compressor stall

The stall phenomenon, which is a breakdown from laminar to turbulent flow, can be steady or cyclic. In the latter case it is generally termed *rotating stall*. It is initiated by the occurrence of local reversals of flow between the main air stream and the boundary layer on the metal surfaces of either the impeller rotor or diffuser, or even in the diffuser scroll. In the impeller, stall is generally caused by too high an angle of incidence of the flow as it approaches the leading edges of the blades.

The mechanism of stall in the diffuser and scroll, on the other hand, is as follows. The thickness of the boundary layer may range from a few microns upwards. Under conditions of laminar flow, the velocity of flow within it rises from virtually zero at the surface until it equals that of the main flow.

Stall can occur in either of only two conditions. One is if the velocity, and therefore kinetic energy, of the layers of the main flow adjacent to the boundary layer falls to the point at which it is inadequate to drag the latter along with it. The velocity gradient between the two layers then becomes locally very steep, so the boundary layer tends to be rolled instead of dragged along the metal surface. The turbulence thus generated becomes thicker as it passes downstream until, ultimately, it can spread rapidly out of control.

The other condition in which stall can occur is when the angle of incidence of the flow as it approaches an impeller blade or guide vane is large and its velocity small. As the angle of attack of the blade to the approaching air flow increases, so also does the distance between the streamline flow and the surface towards the trailing edge of the vane. Ultimately, there comes a point at which the Bernoulli depression generated between the air stream and the surface becomes too small to draw the flow down on to that surface. It therefore breaks away and, as it does so, initiates near the trailing edge turbulence which immediately spreads forwards, over the whole surface.

12.8 Improving compressor efficiency

The higher the pressure ratio, the greater will be the inflow velocity, but increasing pressure ratio increases the tendency to surge. Moreover, high Mach Nos. at the inlet both encourage surge and reduce efficiency. Consequently, because the compressor map is so narrow, Fig. 12.9, a highly turbocharged engine is inevitably operated close to the surge line, especially at part load.

One way of improving efficiency and delaying surge is to increase the number of blades on the rotor. However, in practice, a limit is imposed by the reduction in intake area consequent upon the increase, and a resultant choking of the flow. Increasing the number, therefore, generally entails reducing thickness of the blades and thus impairing their strength and fatigue life. Choking is may be alleviated by shortening alternate blades, by removing from each a portion extending radially outwards from its leading edge.

A device that is used both to improve impeller efficiency and to help to ward off surge is backswept blade tips. One effect of back sweeping is a reduction of the pressure differential across the tips of the blades, and thus a tendency of the flow to reverse locally around them and initiate stall. A similar effect is obtained at the edges of the blades by sweeping them forwards in the case of a compressor and backwards in a turbine, mainly to help to stiffen the blades and induce axial flow in the inlet and outlet respectively. Secondly, but no less important, back sweeping the tips helps to convert some of the large tangential into a radial component of flow, thus modifying its angle as it leaves the tips. Another way of delaying the onset of stall and surge is to position vanes in the inlet, arranging them in a manner such as to pre-swirl the air entering the rotor.

Compressor efficiency can be improved by the introduction of vanes but, owing to an associated reduction in the breadth of the compressor map, over

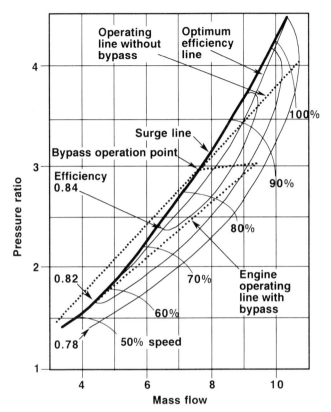

Fig. 12.9 With a turbocharger designed for operation at a high pressure ratio, the performance map is narrow, as shown here. In these circumstances, the engine operating line is inevitably close to the surge line. On this map, the engine operating lines with and without a bypass valve are shown dotted

a significantly narrower range of mass flow. According to Jiang and Whitfield, Proc. IMechE, Vol. 206 D3, 1992, the best results are obtainable with short blades in the diffuser outlet, rather than closer to the tips of the rotor blades, Fig. 12.10. Long blades, extending inwards from the diffuser to the periphery of the rotor are less effective.

A further increase in both overall efficiency and the width of the range can be obtained by varying the geometry compressor diffusers by pivoting the vanes. Various pivoting mechanisms have been used. Some involve fixing to the end remote from the pivot of each vane a peg that projects into a slotted hole in a ring, so that rotation of this ring swings all the vanes about their pivots. Other manufacturers fix the vanes to their pivots, which are rotated by swinging links. Ideally, control is effected electronically, to cater for transient conditions.

The geometry can be varied also by moving one of the side walls of a vaneless nozzle to increase or decrease its width. This can be done either

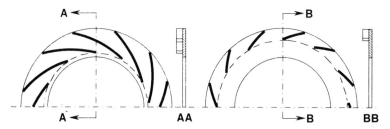

Fig. 12.10 Higher efficiency is obtained when the inner ends of the diffuser vanes are reasonably well clear of the compressor blades, as on the right, rather than as on the left

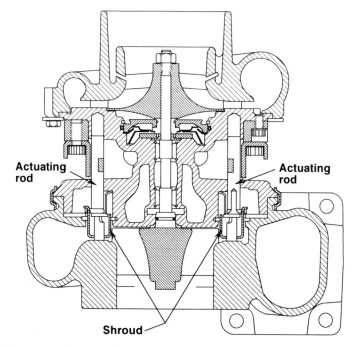

Fig. 12.11 On the Holset variable nozzle turbocharger, a diaphragm type actuator causes the control rods, on one end of which is a shroud, to slide axially. The geometry of the turbine nozzle is thus varied by the movement of the shroud

simply by flexing it, or by pivoting it about one end. For turbines, a device developed by Holset, Fig. 12.11, is a shroud with slots fitting closely around the vanes so that the effective width of the nozzle can be varied by moving the shroud into and out of it. Rods sliding longitudinally in holes in the casing move the shroud. They are actuated by a pivoted yoke which is controlled by either the electronic engine management system or, for gasoline engines, a diaphragm type actuator subjected to manifold depression. Something similar to this might be adapted for compressors, though it would probably be easier to make an aerodynamically efficient flexing or pivoted wall system.

An interesting alternative concept for improving efficiency is the incorporation of a control valve for bypassing part of the compressor output to a point upstream of the turbine. This not only regulates the delivery of air to the engine, but also increases the mass flow through the turbine, which is reflected in higher compressor output, in theory enabling a smaller compressor to be used. The main benefit, however, is that with the bypass valve open at part load, the engine operating line is moved further from the surge line. This is practicable, however, only with constant, as distinct from pulsed, pressure turbine operation, and therefore is of interest primarily for large industrial diesel engines.

A good review of solutions to the problems of compressor design can be gleaned from the previously mentioned Paper by Flaxington and Mahbod, of the Allied-Signal, Garrett Automotive Group.

12.9 Valve timing

Boosting the charge has implications as regards valve timing. First, because turbocharging increases the effective compression ratio, and possibly also the pumping losses, there is a case for reducing this ratio by closing the inlet valve early relative to its timing for natural aspiration. There is no conflict here, but the next two considerations call for compromise favouring the second.

At full load, the boost pressure can be significantly higher than the back pressure from the turbine. Consequently, a large valve overlap can be advantageous because some of the air charge will then pass out through the exhaust, which will lead to an improvement in both scavenging and cooling of critical components such as exhaust valves, combustion chambers and turbine rotor. On the other hand, at low speeds and lighter loads, the boost pressure is likely to be lower than the turbine back pressure leading to back flow of exhaust gas into the induction system. In this case, a smaller valve overlap may be desirable.

These three considerations point to the desirability of using variable valve timing mechanisms on turbocharged diesel engines. Several such mechanisms are either in production or at an advanced stage of development, as described in Chapter 10, though so far they have been applied more to gasoline than diesel engines.

12.10 Turbocharger installation

To take maximum advantage of the pulse energy due to the sudden opening of the valve, the turbocharger should be mounted on or as close as practicable to the exhaust manifold. This has the additional advantage that, if a catalytic converter is fitted, it too can be installed relatively close to the engine and therefore will warm up rapidly to its operating, or *light-up*, temperature.

Ideally, the exhaust gas pressure would be converted into kinetic energy as it leaves the valve port, so that there would be no back pressure to the cylinder, but this is impracticable. Therefore, the gas must be either passed into a plenum and thence at almost constant pressure to the turbine, or directly at fluctuating pressure to the turbine inlet. The latter arrangement, despite the fact that it means that the machine is running continuously at efficiencies that fluctuate with the mass flow, enables best use to be made of the pulse energy to optimise overall efficiency. Consequently, as previously indicated, it is almost invariably the course adopted for automotive applications.

Interference due to overlapping discharge from two exhaust ports into the pipe delivering the gas to the turbocharger tends to reduce the pulse energy at the turbine. The principle is similar to that for inlet valves, Section 11.6. Consequently, on a six cylinder engine, it is desirable to have two manifolds, each receiving the exhaust gas from three cylinders, and delivering it through siamesed ports into what is termed a twin entry volute, Fig. 12.4.

If the turbine is inadequately cooled, a problem can arise. After the engine has been run at high speed and load for a long period and then suddenly stopped, heat tends to soak from the glowing hot turbine into the lubricant remaining in the bearings and casing, perhaps grossly degrading it. Therefore, after working the engine hard over a long period, it is good practice to run it at a fast idle speed for a while, to allow time for both the circulating coolant and the throughput of excess air to reduce the temperature of the turbocharger to a reasonable level before shutting down.

12.11 Turbocharger characteristics and limitations

The exhaust gas driven turbocharger is more compact than most types of positive displacement supercharger. However, the speed at which it drives the compressor, and therefore its effectiveness in boosting the charge, increases approximately as the square of the mass flow through it. Consequently, although it is very effective over the relatively narrow band in the upper part of power range, over which it is matched to the engine, its effect on boost pressure outside that range is insignificant. Indeed, its effect over perhaps the first half of the engine operating range may be almost zero, after which it will rise satisfactorily over the subsequent quarter, and then extremely steeply until the capacity of the passages (usually those through the compressor, though it could be the turbine) is such that it cannot accept any further increase. Installing a larger turbine is not necessarily an option since this will also increase the engine speed at which it begins to become effective.

At the lower limit of effective operation of the compressor, the onset of surge, the frequency of that surge falls with decreasing speed of the engine, and therefore of the rotor, and it becomes zero when the outflow is too small to boost the charge, i.e. as the pressure ratio approaches 1:1.

How the compressor operating range is limited by surge can be seen from Fig. 12.9. In this illustration, the areas enclosed by efficiency contour lines are narrow, and maximum efficiency occurs close to the surge line. To maintain optimum efficiency, therefore, the engine has to be operated near the surge limit. This illustration demonstrates clearly the reason why the turbocharger is effective over such a narrow range of engine operation.

However, narrowness is not such a drawback for diesel engines, with their relatively narrow speed range, as for gasoline power units. The turbocharger is generally matched to the diesel engine in a manner such that it becomes most effective in its region of maximum torque, which may begin at little more than half its maximum speed. Under lighter loads, when the diesel engine is operating with excess air, the fact that the effectiveness of the turbocharger falls steeply with engine speed is not of such serious consequence as regards either overall engine efficiency or exhaust emissions.

12.12 Turbocharger lag

As mentioned in Section 12.1, turbocharger lag is a delay between the depression of the accelerator pedal and the response of the engine, and therefore between the start of acceleration and the time for its completion. Indeed, unless measures are taken to reduce lag, the engine will respond so slowly to transient changes in load and speed, that multi-speed transmissions may be needed for obtaining reasonable driveability. A significant part of this lag is due to the inertia of the turbine and compressor rotors.

The turbine rotor accounts for about 60% of the total inertia which, for obtaining acceptable performance of the unit under transient conditions, needs to be kept to a minimum. Two counter measures can be taken: one is to reduce the number of blades and the other, currently being explored, is to employ a material that is lighter or has a higher strength:weight ratio, especially at elevated temperatures. Examples include ceramic and titanium. A disadvantage of ceramic is that thicker sections have to be used for blades. Consequently, fewer must be used if choking of the outlet is to be avoided. Titanium is of course more costly than steel.

On a diesel engine, however, factors other than turbocharger inertia influence turbocharger lag. In a naturally aspirated diesel engine, all the air needed for complete combustion at full load is present in its cylinders before injection begins. Consequently, full load torque is developed virtually instantly the accelerator pedal is depressed and quantity of fuel injected therefore increased. In a boosted version of the same engine, however, the result of a sudden injection of extra fuel would be black smoke issuing from the exhaust until the air supply caught up. Unfortunately, it cannot even start to catch up until the exhaust gas mass flow into the turbine has increased, to accelerate the turbocharger. Clearly, therefore, any increase in torque in the boosted range can take place only progressively. In this respect, therefore, the diesel engine is at a disadvantage relative to its gasoline counterpart.

12.13 Matching the turbocharger to the engine

The match between turbocharger and engine has to be a compromise, depending upon both the type of vehicle and of operation on which it will be principally engaged. Basically, because of the narrow effective speed range of the turbocharger, if it is matched to the engine at the speed at which it develops maximum power, there may be insufficient air when the speed falls as the load increases, and *vice versa* if it is matched at the speed at which maximum torque is developed. Moreover, matching for peak torque introduces the additional dangers of unacceptably high peak cylinder pressures at maximum speed and, with fixed valve timing, high pumping losses. Another point is that if the turbocharger is matched to the engine at the upper end of its speed and torque range, particulates in the exhaust may become a problem at the lower end.

The design of the turbocharger itself is of course the primary consideration: acceptably high overall efficiency can be obtained only after the efficiencies of both the turbine and compressor have been matched to optimise overall efficiency. A turbine, mainly because it does not have to operate against an adverse pressure gradient, can operate over a wider speed range than its associated compressor, so it is usually the performance of the latter that is critical as regards ease of matching. If the compressor characteristic map is narrow, the decision on how to match it in relation to the engine torque curve is not easy. However, various measures are available for broadening the effective speed range. For example, two turbochargers can be installed in series.

This has been done by KKK with their Turbobox system, Fig. 12.12. At low speeds, one of the turbochargers is bypassed and, as it approaches its strangulation point, the second is brought into operation too. In this case, the point at which the switch is made from one to two units is in the maximum torque range of the engine at medium speeds. An alternative strategy is to use a small turbocharger at the lower end, then switching over to the larger one only and, finally, bringing both into operation, Fig. 12.13.

The principal advantage claimed is good fuel economy. However, the obvious disadvantages are increased bulk, weight, cost, and complexity and and therefore potential for unreliability. Moreover, under transient conditions, switching between modes of operation can cause driveability problems. Whether this can be overcome by appropriate electronic control remains to be seen. The pertinent question is whether a sub-assembly comprising two turbochargers and their electronic control is commercially more acceptable than a costly multi-speed transmission.

The method most commonly used to broaden the range of effectiveness of the turbocharger is to incorporate a *wastegate*, Fig. 12.14. This is basically a relief valve which, opened by either an actuator subject to boost pressure

Fig. 12.12 Two KKK turbochargers, the K14 And K27.2 models, mounted on the manifold of what, in series production, has a single turbocharger matched for best torque. The diaphragm type actuators can be controlled to bring either or both of the turbines simultaneously into operation

or, preferably, under electronic control. With electronic control, the wastegate can be opened as either the boost pressure, maximum cylinder pressure, turbocharger rotor speed or air:fuel ratio exceeds a preset value or, for gasoline engines, when detonation occurs. When the valve opens, it bypasses a proportion of the exhaust gases to a point downstream of the turbine. Wastegated turbochargers are matched at or near maximum torque and, as the engine speed rises from that point, the valve is progressively opened, to reduce the mass flow of gas into the turbocharger. The effect of wastegating is illustrated in Fig. 12.15.

In theory a better solution is to control the power output by, in effect, controlling its size. Basically, for low speed operation, a larger turbine is required than at high speed. However, as previously indicated, the large turbine would over-boost the engine at high speeds and the small one would be unable to supply enough air at low speeds. Each set of the curves in Figs. 12.15 and 12.16 represents, from left to right, the characteristics of turbines of increasing sizes. Given that a compromise has to be accepted, it would appear that acceptable results might be obtained by varying the geometry of the nozzle. The ultimate aim is at maintaining the power input to the compressor and therefore velocity of the air flow through it as the engine speed falls, to avoid surge. However, as indicated in Section 12.4, variable

322 OPTIMISING AIR INDUCTION—TURBOCHARGING

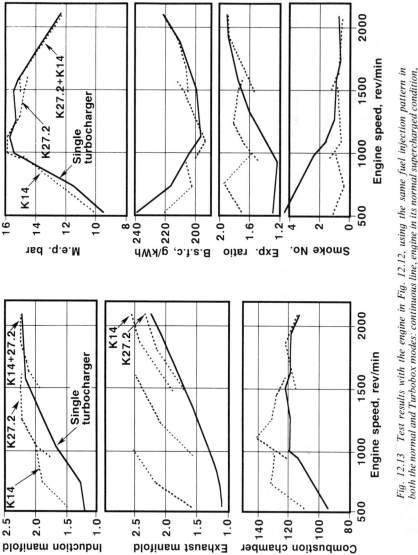

Fig. 12.13 Test results with the engine in Fig. 12.12, using the same fuel injection pattern in both the normal and Turbobox modes: continuous line, engine in its normal supercharged condition, matched for best torque; dotted lines, results obtained sequentially with the K14, K27.2 and, finally, both turbochargers simultaneously in operation

OPTIMISING AIR INDUCTION—TURBOCHARGING

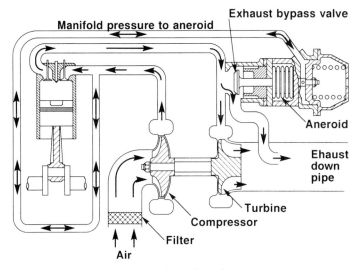

Fig. 12.14 *Diagrammatic representation of an exhaust bypass, or wastegate, system*

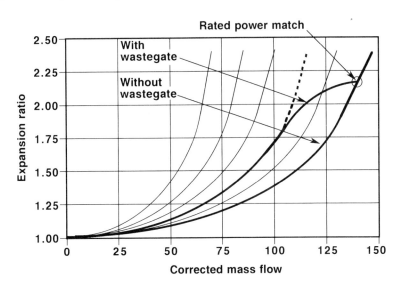

Fig. 12.15 *The six similar curves represent the performance characteristics of a series turbochargers having, from left to right, housings of increasing size. The curves drawn in heavy lines represent the effect of wastegating*

geometry is costly and, especially in view of the wide range of temperatures encountered, an additional potential source of unreliability is introduced.

Of the options mentioned in Section 12.4, the moving sidewall would appear to be potentially the more reliable, though easier in a compressor than at the very high temperatures in a turbine. Pivoting the vanes varies

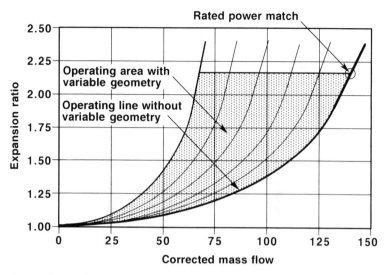

Fig. 12.16 This set of curves is similar to that in Fig. 12.15, but without wastegating. The shaded area shows the effect of varying the geometry of the turbine

not only their angle but also the cross sectional area of the gaps between them, and therefore the swallowing capacity of the turbine. For any given expansion ratio, either of these two methods of varying the geometry produces, over a wide range of operating conditions, an effect similar to changing the size of the turbine, Fig. 12.16. However, since the width of the compressor map, rather than that of the turbine, limits the performance of the equipment, the variable geometry concept is currently gaining ground mainly in the compressor.

Other measures that can be employed include the use of exhaust pipe pulse tuning, on the principles outlined for induction systems in Chapter 11, and using differential gearing to compound the turbocharged engine with a power turbine contributing to the engine output torque. By virtue of the differential gearing, a proper balance between the contributions of the turbocharger and turbine to the total torque output can be obtained under all, including transient, conditions. Compounding therefore offers the potential for greatest benefit but so far has proved to be far too costly, bulky and complex.

A great deal of development work on optimising the performance of turbochargers and overcoming problems relating to the low speed and transient conditions has been done by B.E. Walsham and others of Holset Engineering. The conclusions arrived at have been reported in an outstandingly interesting and informative Paper C405/036 presented in 1990 at the Fourth Intl. Conf. of the IMechE.

12.14 The Comprex pulse charger

The Brown Boveri Comprex pressure wave supercharger suffers neither the matching problems nor the lag associated with turbocharging under transient

OPTIMISING AIR INDUCTION — TURBOCHARGING

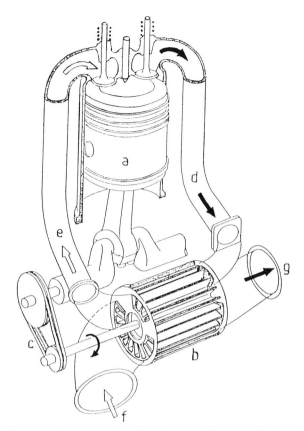

Fig. 12.17 Diagrammatic representation of the Comprex exhaust gas pulse charging mechanism

conditions. Moreover, its output is related directly to mass flow. Exhaust gas energy is utilised to compress the charge, hence the name Comprex. The machine functions on the basis that a pressure wave travelling along a pipe is inverted on reaching an open end, and thus reflected as a suction wave but, at a closed end, is inverted as a pressure wave. Similarly, a suction wave is inverted on reflection from an open end but, from a closed end, it is reflected as a suction wave.

The machine comprises a belt-driven drum within which, arranged longitudinally side-by-side around its axis, is a series of open-ended channels, of approximately trapezoidal-section, extending from end-to-end, Fig. 12.17. The walls separating the channels are in fact of curved section, as viewed in end elevation, to accommodate thermal expansion without distorting the drum. By virtue of a small clearance between the drum and its cylindrical casing, the power consumption (due solely to windage and bearing friction) is between only 1 and 2% of that of the engine output.

For the purpose of explaining the principle of operation, we shall call one end of the casing the exhaust gas end, and the other the induction system end. At the exhaust gas end, there are two ports, the upper one communicating with the high pressure manifold and the lower one with the low pressure pipe to the silencer and tail pipe. Similarly, at the induction system end, the upper port delivers air to boost the pressure in the induction manifold, while the lower one communicates with the air intake in which of course the pressure is lower.

Compression of the ingoing air is effected by exhaust gas pulses generated in the channels, as their ends are opened and closed on traversing inlet and discharge ports. The number of these channels is large enough for the process to be virtually continuous, so synchronisation of rotation of the drum with that of the engine is unnecessary. However, engine performance does vary slightly with speed because the performance of the Comprex machine is related to the length of the channels and the speed of sound at the temperature of the gases in them.

In Fig. 12.18, the ring of channels in the drum is represented as if it were unrolled on to a plane. When the engine is started and the drum rotates, perhaps initially with all its channels filled with air, the right hand end of

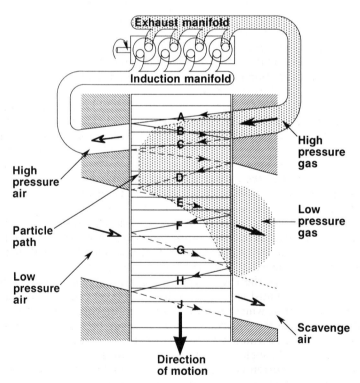

Fig. 12.18 To show the cycle of events of the Comprex pulse charger, the cells are shown here as if they were laid out side-by-side in a plane instead of arranged in a circle within the drum

each channel in turn is opened by its traversing the high pressure exhaust gas port communicating with the manifold. The cycle of operations can be followed by following the events occurring in the first channel to open to the high pressure exhaust port.

A pressure pulse, A, is generated by the sudden opening of this port, and is transmitted at sonic velocity along the channel, compressing the air ahead of it. While it is travelling towards the opposite end, it expels the air that originally filled the channel, through the air port on the left, to boost the pressure in the induction manifold. At the same time, it is drawing exhaust gas in behind it, to replace the air being driven out. As the drum continues to rotate it simultaneously closes both of these ports, to prevent exhaust gas from entering the induction system. The distances, first to the left and then to the right, travelled by the first particle of exhaust gas to enter the channel are represented by the finely dotted line, which is termed the *particle path*, which also represents the position of the boundary between the two working media.

By the time the wave front reached the progressively opening port on the left of the channel, that port would not have opened far, so the area of the opening would have been less than that at the right hand end. Consequently, a compression wave, B, is reflected back to the right-hand end, but not reaching it until the high pressure exhaust port at that end is closed. During this period, the pressure in the exhaust gas driving the wave is lower than that still being discharged into the high pressure air port. Therefore, when wave B reaches the right-hand end, an expansion wave, C, is originated and separates the now motionless and partially expanded gas on the right from that on the left, which is still moving.

By the time this wave C reaches the left-hand end of the channel, the high pressure air port is closed and the contents of the channel are at rest. As can be seen from the diagram, the particle path does not advance as far as the high pressure air port. As the drum rotates further, the low pressure gas channel at the right-hand end opens, allowing the exhaust gas to expand through it, generating the expansion wave D. This wave is propagated at sonic velocity to the left, and reaches the left-hand end when the low pressure air port that end is opening. It therefore causes air to be drawn into the channel, in the wake of wave E, which is reflected back to the right to discharge the exhaust gas that originated the process into low pressure section of the exhaust system. Subsequently, waves are reflected back-and-forth and the discharge into the exhaust system continues but, because of the pressure losses at both ends of the rotor, at a falling rate. During this part of the cycle, first all the exhaust is progressively discharged, and then, in the final run up to the point at which the low pressure exhaust port is closed, excess air is drawn through the channel, by wave J, to scavenge it completely. The whole cycle is then repeated but, in the meantime, other channels have of course been going through the same series of processes.

Development is currently being undertaken on the use of an electronic control and modified porting, to provide exhaust gas recirculation, for reducing NO_x emissions. Also, siren-like noise effects have been reduced or obviated initially by setting the channels at varying pitches and subsequently by installing two rings of channels on concentric differing pitch circles, but offsetting each row by half an acoustical cycle relative to the other. These measures have been reported in Paper C405/032, presented by Mayar *et al*, of AB Turbo Systems, in the previously mentioned IMechE Conference in May 1990 and by Kollbrunner in SAE Paper 800884, in 1980.

Index

Ackroyd-Stuart, 184
Air induction, optimising, 255
Aldehydes, 27, 189
Aniline point, 6
API gravity, 6
Atkinson cycle, 256
Auto-ignition point of diesel fuel, 187
AVL smoke meter, 251

Black smoke, 10, 188, 220
 soiling effect of, 213
Brown Boveri Comprex pulse charger, 324

Carbon, activated, 29
Cars, diesel powered, history, 2
Catalysts suitable for diesel engine emission reduction, 215
Chemiluminescence measurement of HC and NOx, 249
Cetane index, 6
Cetane number, 5
 effects of on ignition delay, 7
 minimum limits of, 7
Cloud point, 7
CNG, 30
 energy density of, 30
 safety of, 32
Cold Filter Plugging Point (CFPP), 7
Cold starting, 190
 aids, 191
 flame heater for, 192
 with direct and indirect injection, 190
Combustion, 184
 chambers, 193
 ignition delay, 5, 188
 inititation of, 187
 Perkins Quadram, 196
 Ricardo Comet IDI chamber, 67
 spray penetration, effect of, 200
 three phases of, 193
Compression, 190
 ratios, 190
 temperature, ultimate in extreme cold, 187
 temperature, ultimate, 187
Compression and spark ignition compared, 187
Comprex pulse charger, Brown Boveri, 324

Diesel cycle, thermodynamic characteristics, 184
Diesel fuels, 1
 influence of quality on emissions, 230
 quality, 13, 187
Diesel index, 6
Diesel knock, 188, 190
Diesel, Rudolf, 184
Direct injection, 193
 comparison with spark ignition, 187, 209
 Rover engines, 311
 small engine design problem, 209
Distributor type pump, 118
 anti-stall stop, Lucas DPC, 138
 anti-stall spring, Lucas DPS, 135
 automatic advance and retard, Lucas DPS, 136
 automatic cold advance override, Lucas DPC, 141
 automatically retarding the timing, Lucas DPC, 141
 axial plunger type, Bosch, 170

330 Index

axial plunger type, definition intro, 118
boost control unit for Lucas DPS, 135
boost controller, Lucas DPC, 148
Bosch single pumping element coaxial with the shaft, 170
Bosch VE, 169
   automatic cold start injection advance, 181
   boost pressure and altitude compensation modules, 179
   combined pump plunger and distributor shaft, 170
   distributor shaft, 170
   fuel metering and distribution, 171
   hydraulically actuated cold start advance, 182
   injection advance and retard, 170
   load-dependent injection timing, 180
   mechanical cold start injection advance, 181
   negative torque control, 177
   positive torque control, 177
   pressure control valve, 169
   pressure (or delivery) valve, 171
   stopping the engine, 183
   temperature dependent starting delivery control, 181
   torque control, 177
   transfer pump, 169
   two-speed governor, 173
Bosch with radially disposed plungers, 173
four plunger rotor Lucas EPIC, 151
Lucas DPA, 119
   all-speed governing, 123
   hydraulic advance/retard mechanism, 122
   hydraulic governing, 123
   hydraulic head, 119
   idling speed control, 124
   metering valve, rotary, 120
   shut-down control, 125
   timing, automatic retardation, 121
   twin opposed plungers, 119
Lucas DPC, 138
   external control of light load advance, 145
   manual light load advance, 143
   maximum fuel adjusting plate, 139
   start-retard, 141
Lucas DPS, 125
   cam box venting, 126
   controlled leakage into cambox, 130
   engine starting, 130
   excess fuel for cold starting, 132
   excess fuel for cold starting, hydraulically actuated mechanism, 132
   fuel supply system, 127
   head locating fitting, or damper, 137
   high pressure delivery valve, 126
   key starting and stopping the engine, 126
   latch valve, 130
   manual idling advance lever, 137
   manual primer, 128
   maximum fuel delivery limiting of, 131
   pressurising valve, 127
   regulating valve, 128
   residual pressure balancing, 126
   rotor vent switch valve, 130
   scroll plates, 135
   solenoid-actuated shut-off valve, 129
   transfer pump, 127
   two-speed governor, 134
Lucas EPIC, 150
   curve of delivery against rotor position, 154
   electronic control, 150
   electronic timing control system, 154
   electronic, hydraulic system of, 155
   electronic, mechanical components, 152
Stanadyne DB2, 158
   cam ring layout, for uneven firing 90 deg V-six, 163
   delivery and snubber valves, 164
   fuel delivery, 164
   fuel delivery ducting, 164
   fuel temperature compensation, 159
   injection timing advance, 165
   maximum fuel adjustment plate, or spring, 160
   pressurising valve, 164
   rotor with tandem plungers for excess fuel control, 162
   snubber valves in the pipe connectors, 165
   torque back-up screw, 163
Stanadyne DS, 166
   angle based control strategy, 169
   DS cam ring advance stepper motor, 166
   DS co-axial spill valve, 166
   DS electronic, 166
   DS electronic control system, 166
   DS housing of the cam rollers and tappets, 166
   DS pump mounted solenoid driver (PMD), 166
Stanadyne Roosa Master, Models A, B, D, DB and DM, 158
Driveability, 28

Electronic control of injection, benefits of, 154
Emissions, 212
   alcohol fuel, with, 26
   aldehydes, 27, 189
   background to, 212
   black smoke, 230
   black smoke, measurement of, 250
   black smoke, sampling of, 244
   carbon monoxide, 220
   CO and CO_2 measurement of, 247
   conflicting requirements for reduction of, 213
   effect of compressed air injection on, 224

Index

effect of insulation of combustion chamber, 221
effect of sulphur content of the fuel, 221
effect of turbocharging on, 218
exhaust gas recirculation (EGR), 217
HC and NO_x, chemilumiscence measurement of, 249
historical review of, 236
hydrocarbons, 218
in different types of operation, 232
legislative control of, 235
limits, 236
   European, 237
   Japanese, 238
   US, 237
oxides of nitrogen, 215
particulates, 220
   and black smoke, test cycles for, 240
   composition of, 220
   definition of, 220
   measurement of, 250
   reduction of, 221
   sizes of, 220
reduction, catalysts for diesel application, 215
sampling of, 244
test cycles, 239
test cycles for heavy commercial vehicles, 239
units of measurement, 235
white smoke, 231
Engine control, 88, 256

Filter Smoke Number (FSN), 252
Filters, fuel, 41
   blocking with solid particles or sludge, 44
   Bosch, 52
   choking with ice or wax, 41, 48
   electric heating of, 48
   for very cold climates, 53
   lives of, 44
   Lucas agglomerator type, 50
   Lucas D-Wax 150 or 300 W units, 52
   practical examples of, 49
   primary and secondary installation, 49
   tests for, 44
Filtration arrangements, 49
Free radicals, 189
Fuel feed pumps, 54
   AC Alpha mechanical, 57
   AC Unitac diaphragm type, 55
   AC Universal Electronic Solenoid type, 59
   Bosch plunger types, 55
   delivery pressures of, 54
Fuel injection equipment, carbonaceous and gummy deposits in, 20
Fuel metering, 78
Fuel system layouts, 42
Fuel tank layout, 36
Fuel, diesel, 1

additives, 13
additives for premium grade, 14
alcohol, 26
alkanes, 6
alpha-methyl naphthalene, 5
alternatives, 25
alternatives, lubrication properties, 26
anti-corrosion additives, 23
anti-foamants, 23
aromatics, 10, 19
ASTM D1 and D2, 3
BS 2869: part 1: 1988, 6
CEN European Standard EN116:1981, 18
cetane number, 5
cetane number and combustion improvers, 15
cloud point, 8, 18, 20
Cold Filter Plugging Point (CFPP), 18
cold flow improvers, 18
compressibility of diesel fuel, 76
corrosion inhibitors, 20
density, 5, 8, 19
detergents, 20
dispersant additives, 20
EEC definition of, 3
effects of constituents, 10
energy densities of various, 30
ethanol, 26
flash point, 5
flow improvers, 18
gaseous alternatives, 28
heptamethyl nonane, 6
hexadecane, 6
Low Temperature Fuel Test (LTFT), 18
lubrication properties, 26
matching additives to, 13
methanol, 26
methanol, cost, weight and bulk of, 28
pour point depressants, 19
properties required for, 5
quality, 3, 13, 230
rate of settling of crystals in, 19
re-odorants, 23
Simulated Filter Plugging Point (SFPP), 18
sources and properties, 5
specifc gravities of various, 28
sulphur in, 5, 212, 220, 221, 230
typical analysis of, 4
vegetable oil derived, 25
viscosity, 5
viscosity, effects of, 9
volatility, 5, 8
water, removal from, 45
wax anti-settling additives (WASA), 18
wax precipitation in, 5, 7, 18
Wax Precipitation Index (WPI), 18
Gas oil, 184
Governing, and associated controls, 88
all-speed, 89
   Bosch electronic, 98

Bosch electro-hydraulic, 100
Bosch mechanical for P Series, 92, 96
excess fuel for cold starting, 85
excess fuel mechanism, tamper-proofing of, 86
functions of, 89
Lucas DPA, 123
Lucas C type, 94
Lucas GE and GX, 99
manifold pressure compensator, Bosch (LDA), 98
matching to suit the requirements of different engines, 95
mechanical, mechanisms, 90
categories of, 89
Minimec, mechanical, 94
speed droop, 90
springs for, 91
two-speed, 89
with torque back-up, 91
with torque control, 89
Glow plugs, 191
Bosch, 192
hot wire type, 191
the ultimate, 192
time to ignition with, 191, 192
Glow tubes, Micronova, 191

Hartford Division of Stanadyne, 158
Hartridge smoke meter, 250
Helmholz resonator, 282

Ignition delay, 188
Indirect injection, 204
combustion chambers, 208
ceramic combustion chambers, strength of, 206
ceramic half-chambers, 206
cylinder capacity, 204
deposition of fuel on combustion chamber walls, 204
direction of fuel spray, 205
disadvantages, 205
high compression ratios with, 205
Nimonic combustion chamber, 206
reducing heat losses, 206
silicon nitride combustion chamber, 208
thermal efficiency of, 205
Induction pipe tuning, 275
Bruntel alternative systems, 297
charge robbery, 284
cluster pipe system, 297
effect of valve overlap, 278, 285
frequencies and wavelengths, 276
Helmholz resonator, 282
Honda Variable Volume Induction System (VVIS), 294
inertia wave, 285
inertia wave, optimising effect of, 285
inter-cylinder interference, 284

Mazda telescopic system, 289
Peugeot Variable Acoustic Supply Characteristics (VASC) system, 292
pipe closed at both ends, 276
pipe closed at one end, 277
pipe end effects, 279
pipe open at both ends, 277
primary pipe cluster and secondary pipe system, 298
resonance, 276
standing waves, harmonics of, 287
standing waves, optimising effect of, 286
telescopic systems, 289
telescopic systems, Tickford, 290
telescopic systems, Mazda, 289
three effects involved, 275
Tickford system, 290
Toyota Acoustic Control Induction System (ACIS), 291
Toyota simple two-stage system, 290
two-stage alternative system, 301
two-stage systems, 290
Vauxhall/Opel Dual Ram system, 290
Volvo siamesed pipe system, 295
zig-zag pipe system, 298
Injection, 73
control of quantity injected, 78
cut-off point, 84
details of spray, 197
dribble and after-injection, 75
dual, 199
on combustion chamber walls, 200
optimum timing of start, 62
lag, 76
pressure diagrams, 68, 69
pressure waves and cavitation, 78
pressure waves, effects of, 64
pressures, general, 61, 62
pressures, maximum, 75
rates of, 75
retard, automatic disengagement of, 85
retarding for starting, 85
spill grooves, 79
spill point, 80
split, 199
spread-over ratio, 78
two-stage, 199
unit, 101
unit, Cummins PT, 113
unit, GM Rochester, 108
unit, Lucas electronic (EUI), 104
unit, Penske/Detroit Diesel electronic, 110
Injection pump, 73
Bosch in-line range, 61
Bosch P Series, 96
in-line, functions of, 74
inline, fuel metering, 78
Lucas Minimec, Majormec and Maximec, 80
performance requirements, 61

Index

plungers, function of, 78, 83
plungers, lubrication, 84
plungers, Minimec, control of, 82
plungers, spill grooves, 79
types of, 73
Injectors, 62
  back-leak, 67
  carbon build-up, 64
  clearance between the pintle and the hole, 64
  fixing of, 62
  functions of, 62
  high spring, 67
  hole type seating arrangements, 65
  low spring, 67
  multi-hole, 200
  nozzle shielding, 65
  nozzles, hole type, 65
  nozzles, Pintaux, 67
  nozzles, pintle type, 63
  nozzles, two-stage, 67
  nozzles, two stage, Perkins, 70
  poppet valve type, 63
  rate of carbon build-up, 65
  sac volume, 66
  Stanadyne Pencil Nozzle, 70
  Stanadyne Pencil, Slim Tip version, 70
  valve opening pressures, 62

LNG, 29
  energy density of, 30
  storage pressure of, 31
LPG, 29
  energy density of, 30
  storage pressure of, 29

MAN M type combustion chamber, 200
Masked valves, 201
Mixing fuel and air, 201
Mixture preparation, 188

Natural gas, 28
  adsorbed (ANG), 29
  adsorbed, storage pressure of, 29
  advantages of, 33
  composition of, 30
  compressed, energy density of, 30
  emissions with, 34
  energy content of, 30
  operational experience with, 34
  reserves of, 34
  safety of, 32
Non-dispersive measurement of CO and CO_2, 247

Oil engine, origins of, 184

Particulate traps, 224
  Deutz-KHD DPFS filter system, 226
  efficiencies, 225
  Ernst Aperatebau, 229
  Iveco, 228
  MAN-Leistritz sequential system, 228
  Mann & Hummel, 229
  Voest-Alpine, 227
  Volvo Cityfilter, 226
  Waschkuttis filter, 227
  Webasto Soot Converter, 228
  Zeuna-Stärker, 227
Particulates, 10
Penske Transportation Corporation Inc, 108
Pierburg smoke meter, 253
Pintles, 63
Pipes, 36
  end fittings and connections, 39
  high pressure, 36
  high pressure, materials of, 37
  multi-layer construction, 38
Plungers, 79
  straight gash spill groove, 83
  alternative spill arrangements, 87
Pour point, 7
Pre-chambers, 204
Primer pumps, manual, 49
Pumping elements, Minimec, 83

Roosa Master pump, Models A, B, D, DB and DM, 158
Rotary distributor type pump, definition, 118

Sedimentors, effectiveness of, 46
Shell Advanced Diesel, 14
Spill port, 76, 80
Signal Series 2000 non-dispersive infra-red (NDIR) analyser, 247
Smoke level units, conversion between FSN and Bosch, 252
Stoke's law, 48
Stoke, 9
Sulphur, effect on catalytic converters, 223
Superchargers, 302
Swirl, 193
  and valve port design, 201
  barrel, 202
  chambers, 204
  generation of, 201
  tumble, 202

Test cycles for particulates and black smoke, 240
Transfer pump, 118
Turbocharging, 302
  aftercooling, 302, 304
  aftercooling, effect on engine efficiency, 303
  bypassing air to upstream of turbine, 317
  compound system, KKK Turbobox, 320
  compressor efficiency, 314
  compressor stall, 313
  compressor surge, 312, 318
  compressor, variable geometry, 315
  compressors, 309

effect on emissions, 218
  exhaust pipe pulse tuning, 324
  installation, 317
  lag, 319
  matching to engine, 320
  nozzles and diffusers, 307
  pressure ratio, 310
  Rover Group engines, 311
  Saab engines, 312
  Scania history of development, 303
  turbocharger unit, the, 318
  turbine eficiency, 307
  turbine rotor inertia, 319
  turbine, variable geometry nozzle, 321
  turbocharger characteristics, 318
  valve timing, 317
  wastegates, 320

Unit injection, 101

Valve timing, variable, 255
  early inlet valve closure (EIVC), 260
  early inlet valve closure, problems, 264
  early or late inlet valve closure, 257
  history, 256
  Honda, 266
  Kolbenschmidt AG, 268
  late inlet valve closure (LIVC), 258
  Mechadyne-Mitchell, 268
  Mechandyne-Mitchell, control of, 271
  simple mechanisms, 265
  two types of, 257
  variable event timing (VET), 267
  variable lift and timing control (VLTC), 266
  variable phase change (VPC), 257, 266
  with throttle control, 264
Vernon Roosa, 70, 118, 158

Water separator, Lucas Waterscan, 47
Water separators, 45
Water, damage to injection pumps, 46
Water, provision for draining, 48
Wax appearance point, 7
Wax, removal of, 48